THE DOOMSDAY DETECTIVES

THE DOOMSDAY DETECTIVES

HOW WALTER AND LUIS ALVAREZ SOLVED THE MYSTERY OF DINOSAUR EXTINCTION

CINDY JENSON-ELLIOTT
ILLUSTRATIONS BY THEO NICOLE LORENZ

TU BOOKS
An imprint of LEE & LOW BOOKS INC.
New York

Copyright © 2025 by Cindy Jenson-Elliott
Illustrations copyright © 2025 by Theo Nicole Lorenz
All rights reserved. No part of this book may be reproduced, transmitted, or stored in an information retrieval system in any form or by any means, electronic, mechanical, photocopying, recording, or otherwise, without written permission from the publisher.

TU BOOKS
An imprint of LEE & LOW BOOKS INC.
381 Park Avenue South, New York, NY 10016
leeandlow.com

Manufactured in the United States of America

Edited by Cheryl Klein and Adam Rau
Book design by Neil Swaab
Book production by The Kids at Our House
The text is set in Calluna, Calluna Sans, Supernett, Pinto, CC Tim Sale, and Indipia

1 3 5 7 9 10 8 6 4 2
First Edition

Cataloging-in-Publication Data can be found at the Library of Congress
ISBN 9781643791098 (hardcover)
ISBN 9781643791111 (ebook)

The facts in the text were accurate and all hyperlinks were live at the time of the book's original publication. The author and publisher do not assume any responsibility for changes made since that time.

To Ronan and Eddie, who are
always up for an adventure
—C. J-E.

For Quin, who asks
excellent silly questions
—THEO

CONTENTS

A Guide to Scientific Practices 1
A DAY LIKE ANY OTHER 3
Chapter One: Earth's Greatest Mystery 15
    Dinosauria 20
    Extinction Is—Impossible? 22
    The Bone Wars Begin 28
    The Dino Disappearance Debate 31
    Too Slow, Too Fast—Two Camps 33
LAWS OF THE LAND: Figuring Out How the Earth Changes over Time 35
    Nicolaus Steno and the Law of Superposition 35
    The View from the Sky 36
    A Clue at the Bottom of the Ocean 40
Chapter Two: The Question of a Lifetime 43
    Geologic Upheavals 52
    In the Heart of the Action 58
    The Question of a Lifetime 60
IT'S ABOUT TIME: How History Is Written in the Rocks 67
    Arthur Holmes and His Geologic Time Scale 74
Chapter Three: A Is for Answers 75
    A Nobel Cause 77
    A Family Affair 81
    Measuring Star Dust 84
IT CAME FROM OUTER SPACE 89

Chapter Four: High-Impact Evidence 95
Supernova 97
A New Idea Every Week 101
All Together Now 103
Nastiest Feud in Science 108

BIG WAVE HUNTERS 111
Clues to an Ancient Mystery 118

Chapter Five: Closing In on a Killer: The Search for the Crater 123
Taking Sides 125
Feathery Ferns of Crystal 129
Shocking Evidence 131
Craters Here, Craters There, Craters Everywhere 132
Signs in the Rocks 134
Sad News 138
Find the Gap 139
Mystery in Mexico 140
Finding a Smoking Gun 144

DOOMSDAY: The Last Day of the Cretaceous 149

Chapter Six: A Never-Ending Story 165
The Heart of Controversy 168
Unlocking the Details 171
Drilling the Impact Site 173
Dinosaur Discoveries 176
The Season of the Impact 182
Big Science, Big History, Big Lives 183

Glossary 188
Source Notes 194
Acknowledgments 203
Endnotes 206
Index 212
Photo Credits 216

A GUIDE TO SCIENTIFIC PRACTICES

As you read, keep an eye out for these key activities scientists perform over and over in any order to answer questions and solve scientific problems.

- **Make Observations:** Scientists use their five senses, sometimes enhanced by tools, to examine the world.

- **Ask Questions:** Scientists wonder about something, and then put together a specific question to express their wonderings.

- **Modeling:** Scientists make and use models to explain how things work. Models change as their understanding changes.

- **Collaborate:** Scientists work together to answer questions and solve problems.

- **Gather Data:** Scientists gather as much information related to the question as they can. The information can be *qualitative* (descriptive) or *quantitative* (using numbers and measurements).

- **Hypothesize:** Scientists create an idea, called a *hypothesis*, that can be tested to explain as much of the existing data as possible.

- **Investigate, Challenge, and Test:** Scientists find ways to test hypotheses using quantitative data.

- **Analyze:** Scientists examine what all the data means, especially the new data that results from testing their hypotheses.

- **Argue Using Evidence:** Scientists present ideas and share evidence to argue why their ideas are correct. Other scientists challenge these arguments.

- **Communicate:** Scientists share their data and ideas with the world.

- **A Note on Measurements:** The world of science, and much of the world in general, uses the metric system to measure everything from mass to length to volume, so this book uses metric measurements. The metric system is based on multiples of ten, so it is easy to remember, since humans have ten fingers and ten toes. Here's a handy chart to help you turn metric measurements into standard American (aka Imperial) measurements.

LENGTH		
1 cm = 0.3937 in.	1 m = 39.37 in.	1 kilometer = 0.62 miles

To change 10 kilometers into miles, for example, multiply 10 kilometers by 0.62.

10 x 0.62 = 6.2 miles = 10 kilometers

A DAY LIKE ANY OTHER

Place: Planet Earth.

Time: The Cretaceous Period, 66 million years ago.

We call the Cretaceous Period "the Age of Dinosaurs," but . . .

The circle of life spun on—

beautiful and tender, terrible and deadly—

a dance of predators and prey on land and . . .

DOOMSDAY.

CHAPTER ONE
EARTH'S GREATEST MYSTERY

Reverend William Buckland gazed over the noisy crowd in the meeting room of the Geological Society of London on February 20, 1824. Outside, the wind whistled down the narrow Covent Garden street. Clouds threatened rain. Inside, the air carried the damp heat of too many bodies in too small a space. In the past few years, the society's membership had swelled to over four hundred men. Tonight, members and their guests were pouring in to see a mysterious treasure hauled all the way from the Devon coast.

This would be Buckland's first meeting as president of the GeolSoc, or Jollsoc, as it was nicknamed. There was nothing he liked more than making a splash. Even though he worked in serious professions as a minister and **geology** professor, Buckland was a bit of a showman. He amazed his friends with outrageous ideas in the name of science. In his home, for example, he was known to allow guinea pigs, a jackal, and a costumed bear to roam at will, while he served his

This picture of a GeolSoc meeting, thought to have been sketched by member Henry De la Beche, shows how members gathered to hear lectures, view artifacts, and refine scientific ideas.

dinner guests hedgehogs, crocodiles, and "toast of mice." He could hardly find a more brilliant way to begin his GeolSoc presidency than with this Devonshire discovery, which would reveal secrets of the ancient world.

A hush fell over the crowd as Buckland's first guest, Reverend William Daniel Conybeare, rose to speak.

Buckland had invited the young minister and geologist to present the fossil, a bizarre treasure discovered and unearthed by a young Englishwoman named Mary Anning near the small seaside town of Lyme Regis. As a woman, and a member of the working class, Anning was not permitted to be a member of the GeolSoc. Social prejudices even prevented her from presenting her own discovery. But she had sent word to her friend Buckland about the fossil almost as soon as she had released it from the rock. He was so excited by the news that he encouraged Conybeare, a colleague of Anning's, to bring it to the society so members could see it for themselves.

Reverend William Daniel Conybeare, 1787–1857.

Conybeare hired a boat to carry the gigantic object from Lyme Regis while he went ahead to London. Rough weather stranded the boat in the English Channel. Conybeare and Buckland had to wait ten days before it was finally delivered to the society at 20 Bedford Street. It took ten workmen to haul it inside, and even then they could not fit it into the meeting room. So it lay on a table in a dim, candlelit passageway while Conybeare presented his talk "On the Discovery of an Almost Perfect Skeleton of the Plesiosaurus."

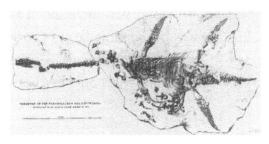

Sketch of a *plesiosaur* from 1823.

Mary Anning's treasure was an enormous fossil frozen in rock. The creature was ten feet long and as wide as a full-grown crocodile, with a backbone and neck that curved like a snake. "To the head of a lizard," Conybeare explained, "is united the teeth of the crocodile, a neck of enormous length, resembling the body of a serpent ... the ribs of a chameleon, and the paddles of a whale." These two-foot-long "paddles" and the long, swan-like neck indicated the creature had lived on or near the surface of the water. It would have been a fine predator of smaller prey, gracefully darting after small fish. Afterward, Conybeare invited geologists to examine the fossil for themselves, by candlelight in the passageway.

When Conybeare was done, he turned the meeting back over to Buckland, who had a surprise of his own to share. He had recently discovered

William Buckland (1784–1856) with fossil finds.

the jaw, teeth, and a few bones of an even larger reptile fossil, which he named *Megalosaurus*.

His booming voice fell to a dramatic hush as he described the *Megalosaurus*—a giant lizard with a thigh bone ten inches in diameter. Even though the *Megalosaurus* fossil was surrounded by the remains of fish and shellfish, it was in the same layer of rock as terrestrial reptiles such as tortoises and crocodilians. Therefore, Buckland concluded, this was an animal that could live both in and out of the water.

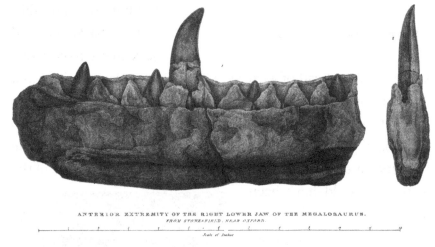

Megalosaurus fossil.

Conybeare and Buckland had successfully delivered the first scientific accounts of dinosaurs ever presented in a public forum. Later, Buckland wrote in high style about the event. "With two monsters of such a kind, and so crowded an audience, my first meeting of taking the chair as President was one of great éclat," a French word that means brilliance and social success.

It would be ten more years before British anatomist Richard Owen would coin the name *Dinosauria* to describe the order of giant reptiles revealed at Buckland's meeting. By then, the dramatic age of dinosaur discovery was well underway.

Mary Anning

Today, Mary Anning's fossils are displayed in museums, and her scientific discoveries are acknowledged as the work of one of the world's most important early paleontologists. During her lifetime, however, Anning lived in poverty, and she rarely attained much scientific credit or financial gain from her work due to gender and class discrimination.

Mary Anning was a pioneer in paleontology. In her lifetime, however, gender discrimination prevented her from fully benefiting from her discoveries.

Born in 1799, Mary Anning began fossil hunting as a child at her father's side, visiting beaches on the southern coast of England and bringing back "curiosities" to sell to tourists. In 1810, when her father died, eleven-year-old Mary and her brother Joseph began supporting the family by fossil hunting, selling their discoveries at a table by the road. A year later, Joseph discovered a giant skull with eye sockets the size of saucers in the Dorset cliff. Twelve-year-old Mary led a team of quarrymen to find and dig up the rest. She spent months putting the fossil—an unnamed seventeen-foot monster—back together, piece by piece, and sold it to the lord of a local manor for twenty-three English pounds, enough money to feed her family for six months.

No one knew what to make of Mary's strange find. Scientists argued about how to **classify** it and what animals it might be related to. It would be ten years before they agreed on a name for the creature, which was eventually displayed in the British Museum in London as an *ichthyosaur*, or "fish lizard." And it was not until December 1823 that Mary Anning, now twenty-four, found the fossil of a second giant sea reptile—the *plesiosaur*, or "near-to-reptile"—and scientists began to wonder whether these creatures might have had even more relatives roaming the ancient world.

Sporting a long, heavy skirt and wearing pattens—raised wooden overshoes—Anning spent a lifetime scaling muddy cliffs in search of relics of the ancient world. Over the years, collectors from around the globe sought her expertise; Buckland, Conybeare, and even the King of Saxony visited her for fossil-hunting adventures or to buy treasures. But while friends published papers about her *ichthyosaurs*, *plesiosaurs*, and *pterosaurs*, the woman who had discovered them remained in poverty. Her friend Anna Pinney wrote, "According to her account, these men of learning have sucked her brains and made a great deal by publishing works, of which she furnished the contents, while she derived none of the advantages."

While Anning was never admitted to the Geological Society because of her gender, she counted its dignitaries as friends, in particular William Buckland and his wife, Mary Morland Buckland. When Anning became ill with breast cancer, fossil enthusiasts paid for her care until her death at age forty-seven in 1847. She was buried in a churchyard in Lyme Regis, and her GeolSoc friends installed a stained-glass window there in her honor.

DINOSAURIA

No one really knows when humans first found dinosaur bones. Perhaps a prehistoric cave dweller found a femur embedded in stone and wondered what it could have been. There are no written records of such an event.

The earliest documentation of a possible dinosaur discovery comes from the Sichuan Province of China, an area of rich fossil beds. In 300 CE, Western Jin Dynasty historian Chang Qu wrote about the discovery of "dragon bones." Dragons held a special place in Chinese mythology. They were considered wise kings of all animals, bringers of rain, and keepers of the pearls of wisdom. As early as 1500 BCE, Shang dynasty artisans had crafted dragons in stone and bronze, and "dragon bones" were an essential ingredient in Chinese medicine. Fossils may have inspired this ancient belief in dragons.

Stories about the dragons depicted in ancient Chinese art may have had their origin in discoveries of dinosaur bones in what is now China.

The English also had creative explanations for the gigantic bones they dug up in British quarries. Around 1677, Dr. Robert Plot, the first Keeper of the Ashmolean Museum in Oxford, wrote about the discovery of a gargantuan femur, or thigh bone. At first, he thought it must be a bone from an elephant brought by invading Romans 1500 years before. But when he compared the femur to the bones of an elephant, he realized it must be a different animal altogether. The femur, he wrote, "has exactly the figure of the lower most part of the Thigh-bone of a Man." Therefore, he decided, it must be a bone from one of the giants mentioned in the Bible.

As more fossils were uncovered, scientists began to notice reptilian qualities in the fossil bones. Some fossils' teeth resembled those of iguanas, for example. Geologists and anatomists debated at GeolSoc meetings about what kind of reptile could have left behind such large bones.

In 1841, Richard Owen, curator of the Hunterian Museum in London, called for an inventory of all the giant fossil reptiles that had been found in England. He hoped to classify, or group, ancient animals with similar features, just as he classified living creatures to see how they were related. When his survey was complete, Owen recognized three distinct groups of ancient reptiles: the meat-eating *Megalosaurus*, the plant-eating *Iguanodon*, and the armored *Hylaeosaurus*. Besides the fact that all three types of animals were no longer

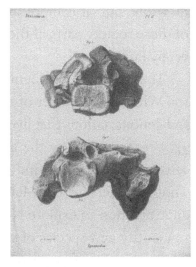

Sketch of *Iguanodon* bones by Richard Owen.

living, Owen noted that unlike any living reptiles, all three had five fused vertebrae in their lower spines. This similarity put them in a class of their own and showed that they were closely related to each other. He named the entire group of extinct reptiles *Dinosauria*, which meant "terrible lizard." Then he asked the question that would drive scientific exploration for the next two centuries: What caused dinosaurs to become extinct?

EXTINCTION IS—IMPOSSIBLE?

Extinction was a new concept in the late eighteenth and early nineteenth centuries. The idea that a species could die out globally seemed impossible. Many people thought that perhaps the missing animals still lived somewhere else on Earth, waiting to be discovered. In 1771, zoologist Thomas Pennant wrote about fossil elephants such as mammoths and mastodons in his book *Synopsis of Quadrupeds*. "As yet the living animal has evaded our search," he wrote, but "it is more than possible that it yet exists in some of those remote parts of the vast new continent, unpenetrated as yet by Europeans."

The "vast new continent" he was referring to was North America. While every corner of the continent was well-known to the Indigenous nations that lived there, this part of the world had not yet been explored by the colonial powers of Europe. European settlers were eager to find out what was there. When President Thomas Jefferson sent Meriwether Lewis and William Clark on an expedition to explore North America in 1804, he asked them to identify new species of plants and animals. They found many species never before seen or recorded by people of European descent, including fossil fish in what is now known as the Hell

Creek Formation. Discoveries like this fueled speculation: perhaps ancient animals still lived somewhere on Earth.

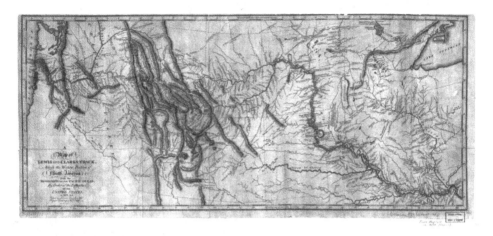

In 1804, President Thomas Jefferson's personal secretary, Meriwether Lewis, and Army officer William Clark, guided by Sacajawea, a skilled interpreter and teenage member of the Shoshone tribal nation, led an exploratory expedition northwest along the Missouri River from St. Louis to the Oregon territory and the Pacific Ocean.

Moreover, to many people in this religious age, the very idea of extinction was blasphemous, meaning against God. Why would God, who created everything on Earth, turn around and destroy what He had made? It was like suggesting that God could make a mistake.

By 1825, however, evidence was mounting that some animals had disappeared from the earth forever. Georges Cuvier, a scientist at Paris's Museum of Natural History, was the first to consider extinction in a serious way. In his book *Discourse on the Revolutionary Upheavals of the Earth*, Cuvier proposed that "revolutions" in nature could cause animals to die out. Cuvier's revolutions were natural disasters that completely changed a region of the earth. While nature is usually calm, Cuvier wrote, sometimes big events, such as catastrophic floods, volcanic eruptions, and massive earthquakes, could cause so many animals to die off that a species could not recover.

Cuvier noted that mammal fossils found near Paris could be evidence of such a flood. Perhaps these revolutions had happened in other places around the world. Cuvier's idea was called **catastrophism** and the people who believed in this viewpoint were **catastrophists**.

Catastrophism appealed to English geologists because many were also clergymen of the Church of England. They practiced "**natural theology**," looking for connections between stories in the Bible and nature. Buckland and his colleagues thought the fossils near Paris could be evidence of the great flood in the Bible story of Noah's ark. Extinct species, including dinosaurs, might have been wiped out in that flood.

Georges Cuvier (1769–1832).

Sir Richard Owen compared the anatomy, or structure, of different animals. He coined the name *Dinosauria* to describe the bones of ancient giant reptiles.

Richard Owen had a different sort of religious explanation for extinction. In 1842, he proposed that God had placed dinosaurs in a particular period on Earth because the air had less oxygen at that time, which must have suited these large animals. Dinosaurs died out, he suggested, because oxygen levels increased. Of course, Owen had no way of knowing what oxygen levels were in the deep past. His idea was pure speculation.

Not every scientist took a religious view of extinction. In 1830, English geologist Charles Lyell wrote *Principles of Geology*, a book about how and why the earth changes over time. Lyell introduced the idea of **uniformitarianism**, the notion that Earth changes slowly through predictable, steady forces: earthquakes, volcanoes, wind, and water. Uniformitarianism was the opposite of catastrophism.

Sir Charles Lyell, 1797–1875.

Lyell argued that extinctions occur in individual species and are caused by changes in the environment. He reasoned that living things are suited to survive in a particular place. When their habitat changes, the species that can't survive die out. He also thought, however, that some extinctions could be reversed. If the environment became suitable to dinosaurs again, for example, perhaps they would return. Extinction, in his mind, didn't mean forever.

Catastrophists and uniformitarians were always arguing. GeolSoc meetings became heated as scientists from both sides presented their arguments and evidence.

Then, in 1859, a new book shook the worlds of science and

religion and added fuel to the intellectual flames. Charles Darwin's book *On the Origin of Species* introduced the notion of **evolution through natural selection**. Darwin's idea was that organisms unsuited to a particular environment would die out over time. Organisms best suited for survival in that environment would live to pass on their traits, or features, to a new generation. Over time, these naturally selected traits would lead to the evolution of new species.

Darwin backed up his idea with lots of evidence. For example, he hypothesized that different species of finches on the Galapagos Islands in South America had evolved from a common ancestor. He argued that over many generations, one species of finch evolved into many, each with a unique beak shape that allowed it to eat the food available on its own isolated island. Evolution through natural selection was a slow process, and it fit well into the slow-and-steady uniformitarian model of geology.

While the debate between catastrophists and uniformitarians could be heated, the idea of evolution turned the argument into an inferno. Darwin's book suggested that nature, not God, was the author of the book of life. This was a blasphemous idea to scientists who believed in natural theology. *On the Origin of Species* was—and still is—one of the world's most influential and controversial books.

Charles Darwin, 1809–1882, changed the world with his book *On the Origin of Species*, published in 1859.

Hypotheses and Scientific Practices

For much of history, scientists used logic and the voice of authority to try to explain what they saw in the natural world. If an idea seemed to make logical sense, they argued, it must be correct. They also believed that the ideas of men who were famous, wealthy, or religious authorities were more correct than those of women, ethnic minorities, or people without wealth. These prejudices made it harder for good ideas to gain an audience.

In 1878, however, American mathematician Charles S. Peirce proposed a new way to make objective, scientific judgments. Peirce and others called this process the "**scientific method**." Scientists now refer to "scientific practices" used in any order to find an accurate answer to a question. (You can see the full list of these practices in the preface on pp. 1–2.)

Charles S. Peirce's way of investigating scientific questions ensures that data is accurate and is not influenced by opinions.

 "Science first begins to be exact when it is quantitatively treated." —Charles Peirce

Borrowing 800-year-old ideas from Muslim scholar Ibn al-Haytham, who used experimentation to discover objective truth, Peirce created a "process of investigation": First, develop a question. Then, answer it with a **hypothesis**. Finally, collect data to test whether the hypothesis is false, over and over again. The more times the hypothesis was *not* proven false, the more likely it was to be true.

For example, suppose you ask the question, "Do all eggs contain baby chickens?" To test it, you could create a hypothesis: "As all eggs contain baby chickens, if I hatch a thousand eggs, then a thousand baby chickens will come out." You would then test your hypothesis by letting a thousand eggs hatch. If a thousand eggs hatch and all baby chickens come out, you can be relatively sure your hypothesis has not been proven false yet. If one of the eggs hatches and a baby *Tyrannosaurus rex* emerges, however, your hypothesis will be proven false.

Chickens, like all birds, are the descendants of dinosaurs.

THE BONE WARS BEGIN

While Darwin's book was shaking scientific thought in England, North America was becoming a hotbed of dinosaur discovery. As railroad employees laid tracks across the country for the transcontinental railroad, the new US Geological Survey (USGS) mapped the expanding nation. Workers discovered rich fossil beds everywhere they looked. The Rocky Mountains and Great Plains proved to be treasure troves of fossil skeletons.

Edward Drinker Cope was one of the earliest dinosaur hunters. Cope started working at the Academy of Natural Sciences of Philadelphia at age eighteen, studying and organizing the museum's reptile bone collection. By his early twenties, he had begun making his own dinosaur discoveries on behalf of the museum in the rich mineral soils of nearby New Jersey. When Cope discovered a nearly complete skeleton of razor-clawed predator *Dryptosaurus aquilunguis,* he caught the attention of Othniel Marsh, the first chair of the new **paleontology** department at Yale College in New Haven, Connecticut.

Wherever and whenever Cope explored for dinosaurs, Marsh would show up. At first, the men were friends. Later, they became rivals. Their competition for the best specimens came to be known as the "Bone Wars" thanks to the underhanded and devious ways they tried to undermine each other. Marsh bribed quarry workers to send him exclusive notice when new fossils were discovered, cutting out Cope. Both sent spies to rival dig sites and spread rumors to ruin the other's reputation. But their rivalry carried a rich reward for science. Between them, they uncovered and named over a hundred new species of dinosaurs, from the long-necked sauropod *Amphicoelias* to species of spiky *Stegosaurus* and

Newspapers carried sensational stories about the discoveries and rivalry of Edward Drinker Cope and Othniel Marsh.

bloodthirsty *Allosaurus*. As America's westward expansion became an engine of scientific exploration, Cope and Marsh's thirty-year rivalry was the vehicle that drove paleontology into the public imagination. It was the first of many dinosaur science rivalries.

On both sides of the Atlantic, dinosaurs became a topic of popular fascination. Dinosaur museum displays sprang up like daisies in a bone yard, first at the Crystal Palace in London in 1853, and then spreading far and wide. With the help of sculptor Benjamin Waterhouse Hawkins, Richard Owen created a public garden of giant prehistoric wonders. Patrons could drink their afternoon tea inside an *Iguanodon* sculpture.

The public poured in to stroll among the giants in a world no one had ever before imagined. Fifteen years later, Hawkins and Cope designed a three-story iron lattice to hold the first mounted dinosaur skeleton, a behemoth *Hadrosaurus foulkii*, at the Philadelphia Academy of Natural Sciences. The exhibit drew enormous crowds of visitors from around the world. Dust generated by gawking dinosaur fans threatened the other fossils on display.

Benjamin Waterhouse Hawkins's dinosaur sculptures—while not accurate by today's standards—fueled the public's fascination.

Over the course of fifteen years, Hawkins lectured and mounted displays at the Smithsonian Museum in Washington, D.C., the American Museum of Natural History in New York City, and at Princeton University in New Jersey. The public was in a frenzy for the prehistoric world.

As the years passed, new media embedded dinosaurs in the public imagination. Sir Arthur Conan Doyle's 1912 book, *The Lost World*, became a 1925 silent movie suggesting that dinosaurs might still be hiding in secret pockets of the planet. In 1938, fishermen off the coast of Madagascar caught a fish with leg-like fins, a species of coelacanth known only from the fossil record. Could this be a "missing link" in the evolution from sea creatures to land animals? If the coelacanth still lived in the deep, unexplored oceans, people reasoned, perhaps dinosaurs still lived somewhere in deep, unexplored jungles. The idea became a popular theme in tabloid newspapers, comic strips, and serial movies.

Winsor McCay's early animated film *Gertie the Dinosaur* gave the public a comical way to view these prehistoric creatures.

THE DINO DISAPPEARANCE DEBATE

While dinosaurs were a hot media topic, the question of extinction remained unresolved. Between the 1920s and the 1960s, scientists developed both serious hypotheses and wild ideas about why dinosaurs vanished. These fell into four basic categories:

1. **Dinosaur Health Issues:** These hypotheses focused on physical weaknesses within the dinosaurs themselves. One scientist suggested that species were not meant to live forever, and eventually, like old tools, they would wear out, due to their excessive weight and expansive appetites. Another scientist argued that dinosaurs' diets must have included ferns, common plants at the time. Perhaps fern oil caused dinosaurs to become chronically constipated, leading to mass death. Another wondered if the pollen from flowering plants led dinosaurs to have fatal fits of hay fever. Without evidence of any of these health issues, these ideas were merely speculation.

2. **Climate Change:** Other scientists hypothesized that changes in Earth's climate were to blame. Perhaps a warming climate had caused the world to become too hot and dry for dinosaurs. Or perhaps, with changes in temperature, the numbers of parasites grew, and dinosaur diseases reached epidemic proportions. Still other scientists wondered if the earth had become too cold for dinosaur eggs to support the life inside them.

3. **Geological Change:** Many scientists argued that changes in the geology of the ancient earth played a role. In North

America and India, active volcanoes spewed lava. Perhaps volcanic activity left the air too warm from lava, too cool from volcanic smoke, or too poisoned from volcanic ash to support dinosaur life.

4. **Astrophysics:** In the latter half of the twentieth century, even astrophysicists jumped into the fray. They suggested that the dinosaurs' extinction may have started in space. Scientists came up with explanations ranging from sunspots to cosmic **radiation**, from interstellar dust clouds to explosions of nearby supernovae.

Everyone wanted to join the discussion. Climate scientists who knew nothing of paleontology, and astrophysicists who knew little about the fossil record all offered opinions on extinction. But without a way to collect data and evidence, it was impossible to prove whether anyone was right or wrong.

Science Word Roots

Words in science often use word roots, or parts, from ancient Latin or Greek. If you know what a word root means, you can often guess the meaning of the whole word. See below for a list of word roots that make up many science words. Put them together! What new words can you make?

Example: geo (earth) + ology (study of) = geology (study of the earth)

geo—earth
saur—lizard
astro—star/space
bio—life
zo/zoo—animals

cosmo—universe
paleo—ancient
bot/botan—plants
phys—nature/natural order

micro—tiny
ology—the study of
ist—a person who does or believes something
ism—a philosophy

TOO SLOW, TOO FAST—TWO CAMPS

In the mid-1960s, scientists began to alter their approach to the question of what caused dinosaur extinctions. They began collecting data about the Cretaceous Period in general, and the extinctions in particular. Paleozoologists looked at changes over time in animal life and dinosaur **physiology**—that is, how the dinosaurs' bodies worked. Paleobotanists collected evidence from fossil plants. Astrogeologists explored records from Earth and space to see if anything from outside the planet might have caused the extinction.

By the 1970s, scientists were gathering into two distinct camps: **gradualists** and catastrophists. **Gradualism** described the gradual decline of dinosaurs in a world that changed slowly. This idea fit into the uniformitarianism views of many geologists. It held that dinosaurs died out slowly, over millions of years, due to climate change, rising sea levels, or other global processes. Catastrophists looked for evidence that the end of the Age of Dinosaurs had come suddenly and decisively, perhaps through a worldwide disaster such as a nearby **supernova**. Each camp had some evidence to support its views. However, finding the origins of events that happened 66 million years earlier was like searching for a single *Iguanodon* tooth in a mountain of petrified toothpicks.

What killed the dinosaurs? By the end of the 1970s, scientists were no closer to answering the question than William Conybeare had been when he spoke at the Geological Society in 1824. But like Mary Anning's giant *Plesiosaur* fossil, left alone in a dark hallway, an answer was just out of reach, waiting for a scientist to widen the doorway and let in some light.

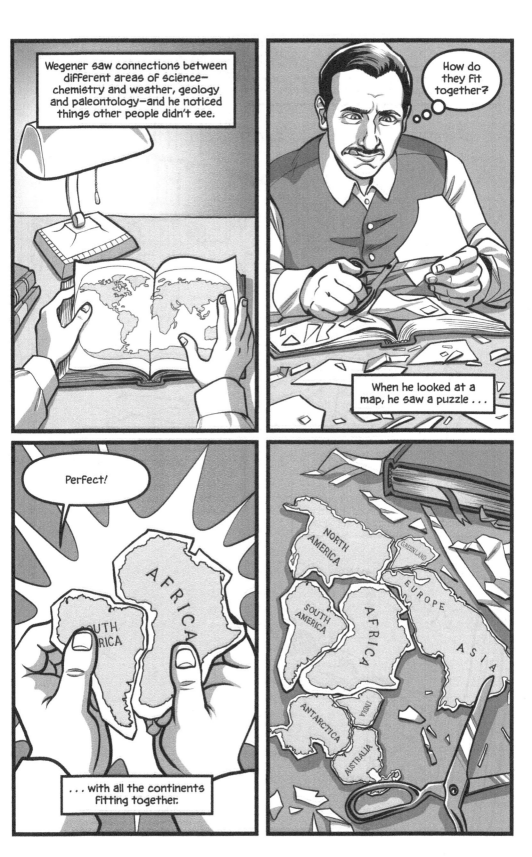

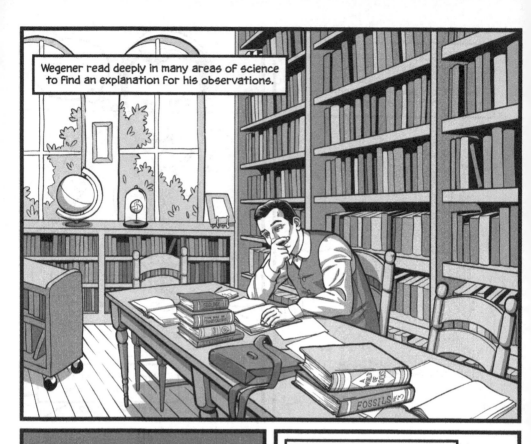

Thirty years after Wegener's death, during World War II, submarine commander and geologist Harry Hess used **sonar** to bounce sound waves off the ocean floor to measure its depth. To his surprise, the ocean floor was not flat, as everyone supposed. Instead, it was a landscape of mountains, canyons, and plains.

After the war, when Hess became a geology professor at Princeton University, the US Navy continued to explore the Atlantic Ocean—including a ridge of high mountains running right down the middle, over 16,000 km long. Harry Hess made this **mid-ocean ridge** the focus of his research.

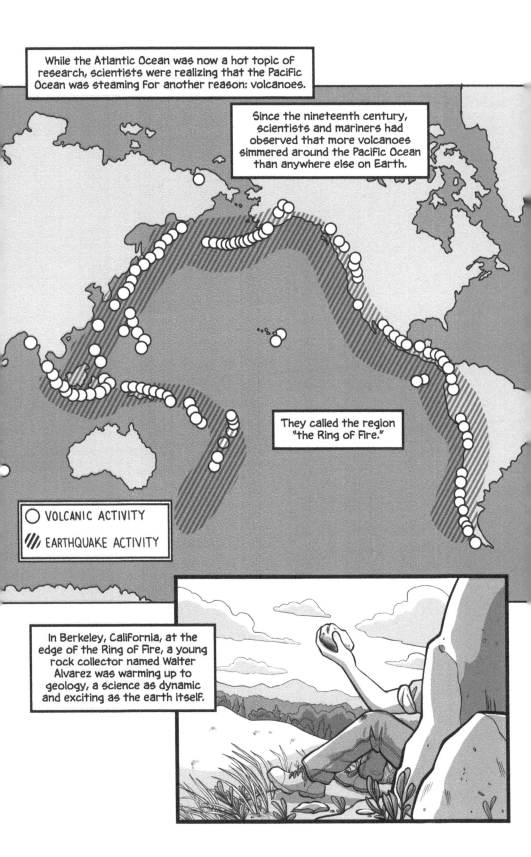

CHAPTER TWO
THE QUESTION OF A LIFETIME

Eleven-year-old **Walter Alvarez was** roaming his Berkeley Hills neighborhood with his friend eight-year-old Gary Messinger, looking for something fun to do. There was nothing the boys loved more than playing pranks, like the time they used soap to paint a sign on the sidewalk that read, "Caution, Apes Crossing."

Suddenly, Walter had a more serious idea.

"Do you want to go see where my dad works sometime?" Walter asked.

"Sure, why not?" Gary answered.

Walter knew Gary had never been inside a scientist's lab, and perhaps had never even visited the University of California, Berkeley, where Walter's father taught. It would be fun showing him the lab and the school where Walter's grandmother had earned her college degree. And, as usual, Walter was curious to see what his father was working on. There was always something new to discover.

Curiosity was an important part of being an Alvarez. It was curiosity that drove Walter's great-grandfather Luis Fernandez Alvarez to move from Spain to Cuba in 1866, when he was thirteen, and then on to California in the 1870s. Luis studied medicine

at Cooper Medical School (later known as Stanford University), then moved to the kingdom of Hawaii in the 1880s to work as a government physician.

He became a leading medical researcher of his time, treating patients with leprosy—at the time an incurable disease—and studying how it affected victims' sight.

It was curiosity that led his son, Walter Clement Alvarez, to develop a lifelong passion for all areas of science. Growing up in the Hawaiian Islands, Walter Clement was fascinated by volcanoes and other geologic events.

Devastation from the 1906 San Francisco earthquake.

From top to bottom: Luis Fernandez Alvarez (1853–1937), Walter Clement Alvarez (1884–1978), Luis Walter Alvarez (1911–1988), and Walter Alvarez (born 1940).

He moved to California to study medicine, and he provided medical aid during the devastating 1906 San Francisco earthquake. After he married Harriet Smyth, a graduate of the University of California, Berkeley, and their children were born, the Alvarez family moved to Mexico, where

Walter Clement worked as a physician for a mining company. He received additional training at Harvard University in Massachusetts, and then the family returned to San Francisco, where Walter conducted physiological research in his lab every morning and ran a medical clinic in the evenings. In 1925, he was hired by one of the world's foremost medical research institutions, the Mayo Clinic in Minnesota, where he taught in the medical school, studied the connections between the body's digestive and nervous systems, and served as the editor of several scientific magazines. After he retired, he began writing books and a newspaper column read by 12 million people, which made medicine understandable to the average person. His work earned him the nickname "America's Family Physician."

Dr. Walter C. Alvarez, "America's Family Doctor," helped the public understand personal health care.

And it was curiosity that led Walter Clement's son, Luis Walter Alvarez, to his own brilliant career in **physics**. Later in life, he said, "All the scientists I know are law-abiding citizens, but they have a healthy skepticism about authority. We are trained to ask 'Why?' continually.... I am convinced that a controlled disrespect for authority is essential to a scientist. All the good experimental physicists I have known have had an intense curiosity that no Keep Out sign could mute."

Walter Clement Alvarez encouraged his son's explorations. As a child, Luis spent many happy mornings at his father's lab in San Francisco, though he was less interested in the human body than he was in the electrical equipment. By the time Luis was eleven, he could measure electrical resistance and make a circuit, and he even built his own crystal radio, an early form of wireless radio, from scratch. When the family moved to Minnesota, Luis attended physics lectures with his father and spent summers as an apprentice in the Mayo Clinic's machine shop, learning how machinists create and tune electrical and medical instruments. When it was time for college, he chose to attend the University of Chicago, first studying chemistry, then, as a junior, falling in love with physics.

Luis decided to pursue a **PhD** in physics at the university, where he built its first Geiger counter to measure radiation from outer space.

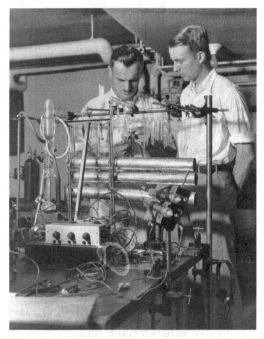
Graduate student Luis Alvarez (right) with his professor, Nobel Prize laureate Arthur Compton, University of Chicago, 1933.

While there, he met and married Geraldine Smithwick, a graduate of the University of Chicago, where she had been a leader in student government and one of the first freshman students to participate in the university's "Great Books" program. After Luis earned a PhD, the couple moved across the country so Luis could work with the famed physicist Ernest Lawrence at the University of California, Berkeley. Their son, Walter, was born in Berkeley in 1940, and their daughter, Jean, exactly four years later, on October 3, 1944.

> "I am convinced that a controlled disrespect for authority is essential to a scientist. All the good experimental physicists I have known have had an intense curiosity that no Keep Out sign could mute."
> —Luis Walter Alvarez

For Luis, as for his father and grandfather, curiosity was a way of life. Later in life, he wrote about how his family valued having an open and curious mind that could connect problems and solutions. "Dad and I talked about the identifying of problems that are really worth working on. . . . He advised me to sit every few months in my reading chair for an entire evening, close my eyes, and try to think of new problems to solve."

A trip to his father's lab always piqued Walter's curiosity. Now his friend Gary would have a chance to see inside too. When the day of the lab visit arrived, Walter and Gary got into Luis's car, and soon they were winding their way around the narrow lanes of Berkeley Hills, through the park-like University of California campus, and past the guard gate to the Radiation Laboratory building, the "Rad Lab," as the scientists fondly called it.

Inside, Dr. Alvarez led the boys to his lab. "If you have watches or anything metal in your pockets, leave them here," he said at the door. The chilly, cavernous space was divided up by crude wooden partitions. Behind one was a giant metal machine. It was the size and color of an elephant, and looked like something you might see in an auto mechanic's shop, but modified for use on a spaceship. Eight feet tall and six feet in diameter, it featured two metal cylinders stacked like layers in a giant cake, each two feet high and separated in the middle by a vacuum chamber full of tubes and wires. A half doughnut of dull gray metal arched over the top, as thick as the stone entryway of a cathedral.

60-inch cyclotron at the University of California Radiation Laboratory, Berkeley, 1939. Luis Alvarez is the second person from the right.

Dr. Alvarez explained that this was a **cyclotron**, a particle accelerator powered by a massive electromagnet. Inside, he told the boys, microscopic particles spin around in circles with magnets, and then they crash into each other, and when the atoms come apart, we can see what's inside. He held up a small steel bar and told them to get out of the way. The boys stepped back, and Dr. Alvarez let go of the bar. It soared across the room, pulled by the powerful electromagnet, and smacked into the cyclotron with a resounding clang.

Next, Dr. Alvarez showed them what the lab had built for visitors: a small glass window four feet high, full of white floating powder. It was a cloud chamber.

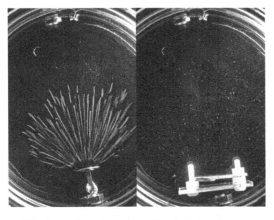

This photo shows how the paths of subatomic particles are made visible in a cloud chamber.

The boys inched closer to the glass. He told them to watch carefully, and they would see the movements that **subatomic particles** made as they were shot through the cloud of powder. He explained that this was how the scientists helped visitors understand things they were not able to see with the human eye.

When they left the lab, Dr. Alvarez opened the door of a long, shabby wooden building. A slim metal machine made of straight tubes ran forty feet along the floor—the entire length of the building—looking like a lumpy steel serpent.

Dr. Alvarez thumped his toe against a tube. "I built this," he said.

Walter looked at Gary. The boy was frowning as if he were unimpressed, wondering why Dr. Alvarez wasn't working on something more interesting. But Walter knew what they were looking at. This was the prototype for a linear accelerator, an atom smasher more advanced than the cyclotron, built for detecting the paths of subatomic particles. This was the machine that took up most of his father's time—as much as eighty hours a week. This was the machine that drove his father's curiosity and ruled his life—the machine that helped him understand the tiniest things in the universe.

Years later, this machine, and others like it, would help Luis Alvarez win the Nobel Prize, the highest award in science.

Walter did not get to spend as much time in his father's lab as his father had gotten to spend at his grandfather's work. For one thing, Luis Alvarez was a busy man, running his lab, working with graduate students, and lecturing all over the world. For another, his father's equipment was expensive—and dangerous. Nuclear physicists couldn't let kids experiment with their tools.

From Walter's earliest days, being in a scientist's family mostly meant moving from place to place to follow research opportunities. Luis once wrote, "Walter was born on October 3 [1940]. We named him for his grandfather. Gerry and I had just moved into a charming house with a spectacular panoramic view of San Francisco Bay. . . . We thought we'd be settling down for a long stay. How inaccurate that scenario was is suggested by the nine different homes Walt knew in the first six years of his life."

Indeed, from 1940 until just after World War II ended in 1945, the Alvarez family followed Luis wherever his work took him— from the freezing winters of Cambridge, Massachusetts, where he developed radar systems to help warplanes fly in bad weather, to the dry heat of Los Alamos, New Mexico, where he worked on the team that developed the atomic bomb.

When Luis's time at Los Alamos concluded, the family settled into an old-fashioned house on a tree-lined street in the Berkeley Hills. Walter was as close to nature as a city boy could hope to be. Broad oaks sheltered meandering lanes; blackberry bushes formed sweet, tangled outcroppings over stone walls; neighborhood parks ended in streams and small waterfalls; and high above the golden hills lay Tilden Park, a preserve with meadows, woodlands, and rushing brooks.

Berkeley was also a city of opportunity. The University of

Luis Alvarez and the Atomic Bomb

By 1940, the United States was deeply concerned that Nazi Germany was close to creating an atomic bomb. The US military recruited scientists to create a nuclear arsenal, **radioactive** "atomic" explosives that could detonate on a massive scale. In 1943, after two years at MIT, Luis was invited to assist Enrico Fermi, a Nobel Prize–winning physicist at the University of Chicago, in purifying graphite for use in a bomb, and assist in developing instruments to fly over Germany to find Nazi bomb-making operations. After six months in Chicago, a researcher named Bob Bacher recruited Luis to work on the Manhattan Project, a top-secret research lab in Los Alamos, New Mexico, dedicated to creating atomic weapons.

At Los Alamos, Luis designed a detonating device for the "Fat Man" plutonium bomb, an atomic bomb made from the radioactive element plutonium. On July 16, 1945, he was on the team in New Mexico that observed the world's first-ever test of a nuclear weapon. On August 9, he flew as a scientific observer in the plane that dropped a plutonium bomb on Nagasaki, Japan, the second of two atomic weapons used on Japan during the war. This bomb killed 80,000 people and drove Japan to surrender, marking the end of World War II. Decades later, in his autobiography, *Alvarez: Adventures of a Physicist,* Luis expressed his sorrow at the terrible loss of life he had witnessed. However, he also wrote: "I believe that the present stability of the world rests primarily on the existence of nuclear weapons, a Pandora's box I helped open with my tritium work and at Los Alamos. . . . I am optimistic enough to believe that the next scourge to disappear will be large-scale warfare—killed by the existence and nonuse of nuclear weapons."

The fireball from the Trinity Test nuclear explosion.

California campus was home to science and history exhibits as well as sporting events, and the community was a mecca for writers, musicians, and artists. Walter's family had neighbors in every academic discipline, from librarians to historians, from lawyers to scientists to businesspeople. To live in Berkeley was to live at the center of civilization.

In the summer, Walter's mother, Geraldine, took him and his sister, Jean, on railroad adventures throughout the western United States. Walter reveled in the spectacular scenery, from volcanic

The Campanile on the UC Berkeley campus has been a landmark since 1915.

peaks to glittering glacial valleys to eroded red-rock canyons. In wilderness areas and National Parks, the family observed the large-scale evidence of a changing planet.

Back at home in Berkeley, Geraldine demonstrated for Walter where and how to collect rocks and minerals, and they spent hours "rock hounding" in the Berkeley Hills. She lent him her rock hammer, a small, square hammer with a flat head for splitting open rocks and a pointed top for prying rocks out of the ground. Walter carried it faithfully on hikes until he lost it one day and had to return home empty-handed.

Walter remained interested in science. He met plenty of scientists and visited many labs, watching investigators work with ideas and instruments. But his father studied ideas so big—and particles so small—that most people couldn't understand them at all. Even with his family's legacy of research, by the time he was in high school, Walter suspected lab work wasn't for him. He wanted to be outside, exploring nature, having adventures in the sun. Later in life, Walter said, "I like being out in the mountains, up in the fresh air and the sunshine with the breeze blowing . . . so geology appealed to me." He decided to study geology, the history of the earth as recorded in its rocks, as real and solid as the planet itself.

GEOLOGIC UPHEAVALS

After Walter graduated from high school in 1958, his curiosity led him to Carleton College in Minnesota, which was known for its geology department.

The science of geology was experiencing a major upheaval. In 1953, American geophysicists Maurice Ewing and Bruce Heezen and geological **cartographer** Marie Tharp had discovered and mapped a deep canyon called "the Great Global Rift" running through the middle of the Atlantic mid-ocean ridge. The **rift**, and other valleys like it, divided the surface of the globe into giant pieces of **crust**. The scientists called these chunks of crust "tectonic plates," and the study of their movement "**plate tectonics**." Plate tectonics could explain how the continents moved, the mechanism behind Alfred Wegener's idea of continental drift, why earthquakes and volcanoes occurred, and how mountains were made. The theory also fit perfectly into uniformitarianism, the idea that the earth changes gradually over time in consistent ways.

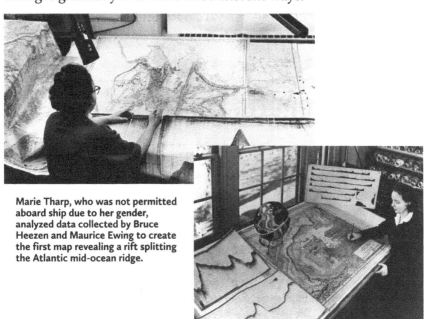

Marie Tharp, who was not permitted aboard ship due to her gender, analyzed data collected by Bruce Heezen and Maurice Ewing to create the first map revealing a rift splitting the Atlantic mid-ocean ridge.

Princeton professor Harry Hess hypothesized that the Great Global Rift produced **magma**, the molten rock found in volcanoes. As magma spewed out, it formed new crust that pushed the plates in the ocean floor apart. Hess called this process "**seafloor spreading**" and believed it was the engine behind the movement of tectonic plates. He hypothesized that if his theory was correct, then rocks farther away from the rift would be older than rocks closer to the rift, and temperatures near the rift would be warmer than those farther away. Later research proved him right on both counts.

That same year, two British geologists discovered surprising evidence that further confirmed Hess's hypothesis. Fred Vine and Drummond Matthews took magnetic readings across the ocean floor and discovered that as they moved farther away from the mid-ocean ridge in either direction, the **magnetic polarity** of the rocks reversed back and forth in wide stripes. This was evidence that the rock was pushing outward over millions of years.

Magnetic Polarity

A compass is a useful tool for finding your way. When you hold it flat in your hand, the arrow will naturally spin to point north. This indicates polarity—the direction of the earth's magnetic field. Currently, the earth's magnetic polarity causes compasses to point north. But if you time-traveled back 780,000 years, the compass would spin around and point south instead. That's because Earth's polarity switches direction, shifting from north to south and back again. Reversals in polarity can occur over periods as short as 7,000 years, or as long as millions of years. The last time Earth's polarity changed was 780,000 years ago. During periods when the magnetic pole is in the north, magnetic crystals in rock align themselves to point north. During other eons, when the polarity shifts and the magnetic pole is to the south, the rock crystals point to the south. By tracking these patterns of **magnetic reversals** in the rock over the years, scientists have been able to create a rough timeline of the switches in polarity. This helps them figure out a timeline of geologic events.

But even as scientists discovered how new crust formed, they still didn't know what happened to old crust. If the earth only made new crust and never recycled the old, the surface of the planet would be getting bigger and bigger. Since that wasn't happening, somehow, somewhere, old crust was being recycled and remelted into magma.

Finally, an accidental discovery revealed how and where the earth's crust is transformed. In 1963, the United States, the USSR (now known as Russia), and the United Kingdom signed a treaty that declared no country was allowed to test nuclear weapons. Because nuclear weapons detonate with such violent force, scientists set up sensors around the world to watch for shaking from any explosions that might violate the treaty. But the sensors also recorded earthquakes—lots and lots of them—almost solely at the edges of tectonic plates. Scientists hypothesized that earthquakes happen when the edge of one plate is pushed against and under another plate. At plate edges, called **subduction zones**, Earth's crust melts and is recycled into the **mantle**, the hot, semi-liquid layer on which the plates float. In 1974, scientists in the *Alvin* undersea vehicle observed volcanic activity in mid-ocean ridges and rifts, supporting Hess's plate tectonics hypothesis. Finally, scientists could pinpoint where new crust was made and what pushed plates apart.

The *Alvin* submersible vehicle recorded evidence of magma at the mid-ocean ridge.

Plate Tectonics

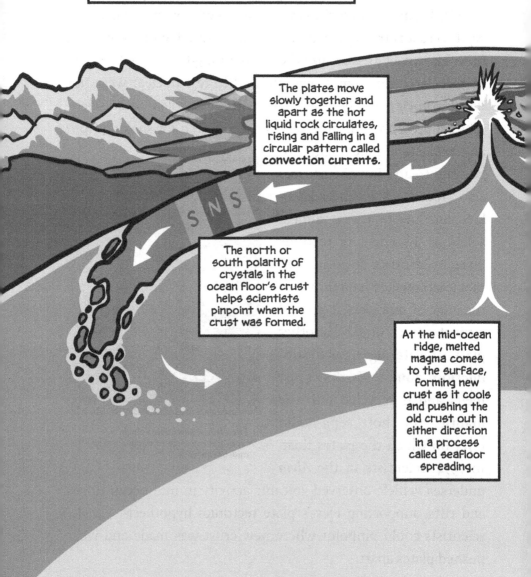

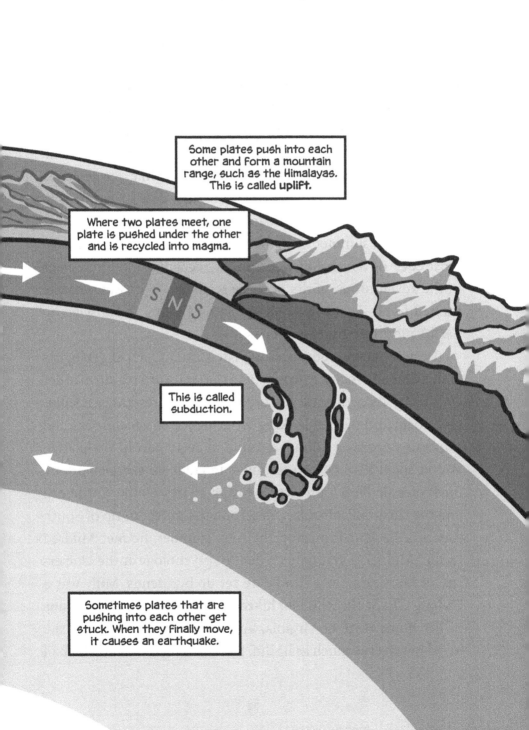

IN THE HEART OF THE ACTION

Studying the new science of plate tectonics ignited Walter's passion for geology. He wanted to be at the center of the storm, in the heart of the action. After earning his bachelor's degree in geology at Carleton College, he decided to attend graduate school at Princeton University and study with Professor Hess.

From his very first day at Princeton, Walter knew he was in the right place. He listened with bated breath while one geologist spoke of driving Professor Hess in a broken-down jeep through parasite-infested rivers in roaring rain in Haiti, and another recounted his summer mapping the wilderness on the northernmost edge of South America. Walter recognized kindred souls in the adventurous spirits of these scientists. This was what he wanted to do—work in far-flung places, finding the answers to Earth's most intriguing mysteries.

Despite the new advances in plate tectonics, most geologists at this time still focused on mapping the strata of a region in order to identify its geological history and natural resources. Geologists roamed the earth looking for evidence of which rocks and minerals were layered underground in every part of the planet. Walter spent his first two summers in graduate school working on Professor Hess's projects creating charts of the Caribbean and the Guajira Peninsula, a desert on the northern tip of South America. Back at Princeton the next summer, he met Mildred "Milly" Millner, a graduate student in psychology at the University of Maryland, on a blind date set up by friends. Milly was a lifelong Girl Scout who had hiked in the Blue Ridge Mountains in her home state of Virginia, and Walter quickly realized she loved to travel as much as he did. She was the companion he had been looking for, a partner for a life of adventure. After they married, they went to the Guajira Peninsula for their honeymoon. Walter's work there led to his PhD in geology, as well as a geological map of the region.

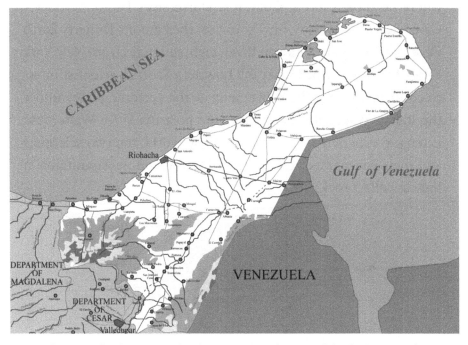

Walter and Milly Alvarez spent their honeymoon traveling around the Guajira Peninsula.

Over the next decade, Walter and Milly had many adventures. They lived in the Netherlands, where he worked for an oil company; in Libya during a revolution; and in Italy, where Walter became interested in archaeological geology, the study of how geology affects human civilization. And yet, for all his research, what Walter wanted most of all was to find an important question that needed to be answered, an exciting question to ignite his curiosity, a fundamental question with an answer that would change the world. He was looking for the question of a lifetime, and he was about to get his wish.

THE QUESTION OF A LIFETIME

In the early 1970s, Walter and Milly returned to the United States, where he took a job at Columbia University's Lamont-Doherty Geological Observatory, now known as the Lamont-Doherty Earth Observatory, on the Hudson River, eighteen miles from New York City. At Lamont, Walter met Bill Lowrie, a Scottish scientist interested in measuring the movement of the tectonic plates under the Italian Peninsula, using magnetic polarity. Since Walter had extensive experience in that part of the world, the two scientists and their wives traveled together to the Apennine Mountains in Italy to collect rocks.

Much of the Apennine Mountains are composed of Scaglia rossa limestone, uplifted from the ocean floor.

Out in the open air near the Italian city of Gubbio, the four friends trooped along winding mountain trails, through sunshine and rain, collecting samples of the Apennines' signature rock—Scaglia rossa limestone. The red color of the Scaglia rossa indicated that it contained rust, or oxidized iron. Since rocks with iron are a good place to look for a record of magnetic reversals, they collected Scaglia rossa from many different layers of limestone, hoping it would give useful magnetic data as evidence of how and when small plates in the region's crust had moved.

When Walter and Bill returned to the United States with their cache of red rocks, they looked carefully at the magnetic traces in the Scaglia rossa. They were disappointed to realize that because of the upward movement of the earth's crust, which thrust the limestone from the ocean floor up into the mountains, the limestone had twisted and turned over time, making it impossible to tell when and how the Italian Peninsula had rotated. They could not put a time stamp on the layers of rocks. They would need a new tactic.

Scaglia rossa limestone is made of the microscopic shells of foraminifera. Its red color comes from iron.

The next year, Walter and Bill Lowrie returned to the Apennines with a group of scientists from Princeton University and a new idea to explore. Instead of looking at the iron in the limestone, they would look at what the limestone itself was made of: the microscopic shells of **foraminifera,** or **forams**—different species of tiny organisms that have floated in the ocean and fallen to the seafloor for millions of years.

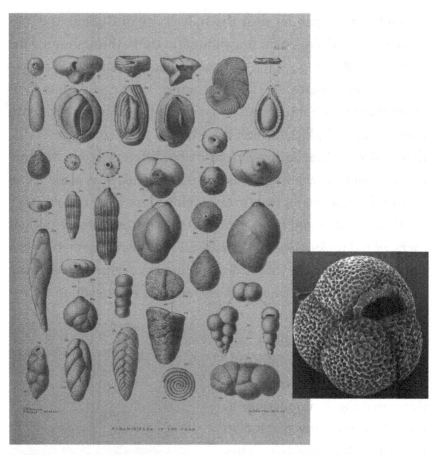

The presence of foraminifera in rock is an indication that the rock once lay at the bottom of the ocean. The species of forams tells scientists when the rock was formed since different species lived at different times.

In Italy, they met Isabella Premoli Silva, an expert on ancient foraminifera.

Premoli Silva could identify individual species of forams from different periods of geologic time from their silhouettes in the rock. With her guidance, Walter and Bill matched forams to magnetic reversals, reading the history of the rocks to figure out when the rocks had formed, turned, and twisted. Their work resulted in a peer-reviewed scientific paper with an impressive name: "One Hundred Million Years of Geomagnetic Polarity History."

The Peer-Review Process

The **peer-review process** is a rigorous way of ensuring that a study is high quality. After scientists use the scientific process to conduct a study, they write a paper describing their experiment and the results. When they submit the paper to a scientific journal for publication, fellow experts in the field, or *peers*, will read the paper to look for evidence that every step of the experiment was done correctly. These experts examine the scientists' process, review their calculations and data, and question their conclusions and analysis. If the experts have questions about the methods used in the study, the steps taken in an experiment, or the results and claims the paper is making, the article is sent back to the original scientists for explanation or revision. The peer-review process guarantees that studies are held to a standard of excellence before they are accepted for publication.

Articles that are not peer reviewed may have errors in their facts, methods, or conclusions, and may actually include information that is incorrect and harmful. This has happened in the past with disastrous results. For example, when inaccurate and falsified articles published in non-peer-reviewed journals have made false claims that immunizations are unsafe or that tobacco use does not cause cancer, this misinformation led some people to avoid immunizations that are beneficial and to believe smoking is safe. The peer-review process ensures that the public can trust the scientific process to find the truth, and trust scientists to report accurate information.

As Walter studied the region's geologic record, one thing struck him as especially strange. While most limestone layers looked the same, two layers, right next to each other, were very different. Limestone near the top layer of the Cretaceous Period (from 145 to 66 million years old) was full of forams of all different shapes, sizes, and species. This indicated that the Cretaceous ocean had been rich with diverse microscopic life. The limestone layer above it, however, from the beginning of the Tertiary Period (from 66 to 2.58 million years old) contained very different forams. They were small in size and much less diverse. This meant that most species of forams alive in the Cretaceous Period had vanished without a trace during the Tertiary Period.

Even more odd was the fact that the two layers of limestone were separated by a thin line of clay, one centimeter thick, containing no fossils at all. The place these two layers met was known as the **K-T boundary**, with the *K* coming from the German spelling of "Cretaceous," and the *T* meaning "Tertiary."* The difference between the two layers, and the strange layer of clay, made Walter curious, especially when he recalled a mysterious and unexplained extinction that marked the end of the Cretaceous.

Scaglia rossa limestone at the K-T boundary.

Walter and Bill continued to travel throughout the Apennine Mountains, uncovering different outcrops of the Scaglia rossa and gathering evidence of forams and magnetic reversals. Everywhere they went, they saw the

---

* In 2008, the name of the Tertiary Period (T) changed to the Paleogene (Pg), and the K-T boundary became known as the **K-Pg boundary**. For consistency, we will refer to it as the K-T boundary throughout this book, until the final chapter, when it will be called the K-Pg boundary.

same thing: a layer of Cretaceous limestone full of large and diverse forams; a layer of Tertiary limestone with small forams; and between them, a thin layer of gray, fossil-free clay. When he returned to the United States, Walter talked to an expert in marine microfossils who confirmed what he suspected. The disappearance of so many forams in the layers of Scaglia rossa limestone seemed to have happened at the exact same time as the most famous extinction of all: the end of the dinosaurs.

Puzzling over the geologic record, Walter took his grandfather's advice and allowed his mind to wander and wonder. Being in the open air and nature had always helped him work out problems, so he took a long walk around the grassy, wooded grounds of the Lamont lab. What happened to the forams? What killed the dinosaurs? Could it have been the same thing?

Years later, Walter wrote, "The question of the K-T extinction looked like one that could lead in totally new directions, and by the time I finished my walk, I had decided that I would try to solve it."

Now Walter had what he had been looking for: a big question that mattered. It was a question that needed answering. It was a question that could drive his science. It was a question that could change our understanding of the world. He had found the question of a lifetime: What caused the K-T extinction?

> "Occasionally there is a question that offers an opportunity for a really major discovery. Choosing what problems and what kind of problems to work on is a critical strategic decision for a scientist."
> —Walter Alvarez

In the 1830s, some members of the GeolSoc became obsessed with figuring out how strata related to the age of the earth.

They noticed that certain fossils were always found in the same layers of rock. Perhaps fossils could help identify the age of strata.

In 1838, William Buckland published a **stratigraphic** wall map showing what fossils were found within which layer. He labeled the layers by the rocks the fossils were found in, such as greywacke and quartzite.

But *exactly* how old was each fossil zone? In the late 1800s, Yale University professor Bertram Boltwood discovered a way to measure the age of rocks that contained uranium.

The same element can take different forms, called **isotopes**, depending on how many neutrons are in its nucleus.

In certain elements, like uranium, the nucleus of an atom can be proton-rich, with extra protons, or neutron-rich, with extra neutrons. Because of these extra subatomic particles, the nucleus goes through a process called radioactive decay, emitting, or ejecting, particles until it is stable—has an equal number of protons and neutrons. If an element emits particles, it is said to be radioactive, and emitting radiation. The time it takes for half of a sample of a radioactive element to become stable is called its **half-life**.

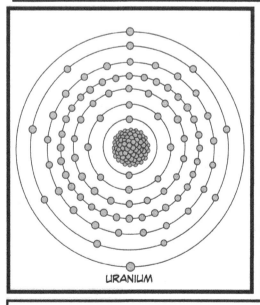

URANIUM

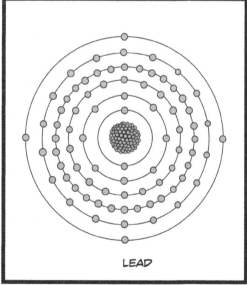

LEAD

As uranium decays, it eventually becomes lead, a non-radioactive element. The time it takes half of a sample of Uranium-238 to decay—its half-life—is 4.47 billion years. By measuring the ratio of uranium to lead in a rock, Boltwood could figure out the approximate age of rocks. He calculated that the oldest rocks on Earth were at least 2.2 billion years old. This process is called radiometric dating.

Arthur Holmes and His Geologic Time Scale

Finally, in 1913, British geologist Arthur Holmes developed the Geologic Time Scale to help organize the billions of years of Earth's history into distinct intervals. Each time marker is a subdivision of the scale before it in the same way that a second is part of a minute, which is part of an hour, which is part of a day.

Eons are half a billion years or more.

Eras, subdivisions of eons, are many hundreds of millions of years.

Periods, subdivisions of eras, are hundreds of millions of years.

Epochs, subdivisions of periods, are tens of millions of years.

EON	ERA	PERIOD	EPOCH
PHANEROZOIC	CENOZOIC	QUATERNARY	HOLOCENE
			PLIOCENE
			MIOCENE
		TERTIARY	OLIGOCENE
			EOCENE
			PALEOCENE
	MESOZOIC	CRETACEOUS	
		JURASSIC	← COMMONLY DIVIDED INTO EARLY, MIDDLE, & LATE
		PERMIAN	
	PALEOZOIC	PENNSYLVANIAN	
		MISSISSIPPIAN	←
		DAVONIAN	
		SILURIAN	
		ORDOVICIAN	
		CAMBRIAN	
PRECAMBRIAN	PROTEROZOIC ← COMMONLY DIVIDED INTO EARLY, MIDDLE, & LATE	NEOPROTEROZOIC	
		MESOPROTEROZOIC	
	ARCHEAN ←	PALEOPROTEROZOIC	
	HADEAN		

Ages are hundreds of thousands to millions of years.

The oldest interval, the Hadean Eon, happened 4.5–4 billion years ago.

Dinosaurs lived from 230 mya in the Triassic Period to the end of the Cretaceous Period, approximately 66 mya.

We are currently in the Phanerozoic Eon, the Cenozoic Era, the Quaternary Period, the Holocene Epoch, and the Meghalayan Age.

Meanwhile, in the Holocene Epoch...

How much time passed as this clay was deposited at the K-T boundary? A thousand years? A million years?

CHAPTER THREE
A IS FOR ANSWERS

By the time Walter settled on his big question, much had changed for his father, Luis Alvarez. In 1958, when Walter was in his last year of high school, Luis and Geraldine Alvarez divorced. By the time Walter moved on to graduate school four years later, Luis had remarried and was starting a new family in Berkeley with his second wife, Jan, an administrator at the Radiation Laboratory. In the next six years, while Walter earned his PhD, married Milly, and pursued his first professional jobs as a geologist, Luis remained at Berkeley. He worked in the Rad Lab as a physicist and raised his two young children with Jan.

In the lab, Luis mentored a team of young scientists and PhD students known as the A Group. The A stood for Alvarez. They dove deep into particle physics—that is, the study of the invisible subatomic particles that make up atoms. The researchers used high-speed

Cosmic ray detection experiments by the Berkeley A Group.

Cosmic Rays

Victor Hess rose high above the earth to measure ionization in the atmosphere.

In 1912, physicist Victor Hess hoisted a measuring tool called an **electroscope** over 15,000 feet up in a passenger balloon to measure ionization in the atmosphere. Ionization is the process by which atoms gain or lose electrons and become electrically charged. Hess knew ionization could be caused by **ionizing radiation**, where radioactive materials emit particles that cause other atoms to lose electrons. Like most scientists at that time, he believed that radiation came from radioactive rocks in the ground. He presumed that the farther he was from the earth, the lower the level of radiation and ionization would be. Hess measured the ionization high above Earth, then compared his data to measurements made on the ground. To his surprise, he discovered that the higher he went, the more electrical charge existed. In fact, the rate of ionization was three times higher 5,300 meters up in the air than it was at sea level! Hess didn't know it, but he had just discovered **cosmic rays**, or radiation from outer space.

Measuring and classifying cosmic rays opened the door to the new science of astrophysics, the study of the chemistry and physics of space. In the 1930s, when Luis Alvarez was a PhD student at the University of Chicago, scientists were aware that atoms were composed of protons, neutrons, and electrons, and they theorized that each of these subatomic particles might be composed of even smaller particles, which might be found in cosmic rays. Luis and his colleagues collected evidence of cosmic rays all over the planet, at different altitudes, latitudes, and longitudes. They set up complex detectors on top of a hotel in Mexico City, on the deck of a research vessel sailing to Peru, and on an airplane flying over Northern Canada. They wanted to know how often cosmic rays enter the atmosphere, and what happens when they do. By the 1960s, Luis and his colleagues had discovered that cosmic rays bombard Earth's atmosphere constantly, colliding with atoms to create a spray of secondary subatomic particles, and emitting electromagnetic energy, such as X-rays, down on Earth.

Luis Alvarez used a wheelbarrow to change the direction of his cosmic ray detector from east to west every half hour on top of a hotel in Mexico City.

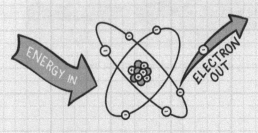

When high-energy subatomic particles from radioactive materials hit atoms, they knock electrons off, turning them into ions. This is called ionizing radiation.

equipment to propel these particles through a liquid-hydrogen chamber to understand their behavior. They also mapped the paths of cosmic rays—particles that fell to Earth from outer space. Luis was known as a demanding teacher and supportive leader, and scientists in the A Group knew they could ask him anything and he would encourage them to explore.

Luis Alvarez (at left) and Andrew Buffington, a young A Group scientist.

A NOBEL CAUSE

In 1968, Luis's professional life changed in a dramatic fashion. On October 30, at 3:00 a.m., a jangling phone startled him from sleep. It was CBS News in New York. A report had just come in from Stockholm, Sweden, that Luis had been awarded the 1968 Nobel Prize in Physics.

After Luis hung up, he and Jan sat together, breathless with the news. Immediately, the phone rang again. It was United Press International, another news organization. When Luis was finally off the phone, the couple talked. The Nobel Prize, they knew, could be both a blessing and a challenge. They had seen how the award had drastically altered the lives of friends, and they wanted to avoid the negative consequences of fame. They decided then and there that their lives would not change in one important way: their present relationships would remain strong. They wouldn't accept social invitations that were given only because of the prize. They valued their real friends, family, colleagues, and research too much to chase after glory.

For the next few hours, the phone rang off the hook. The first call was from Luis's daughter, Jean, phoning from Massachusetts, where she was an instructor at Wellesley College. (Walter, who was working as an oil geologist in Libya, had no access to news or a phone that morning, and only heard the news later.) By 5:30 that morning, photographers from news agencies showed up at their front door. Later that day, the official telegram from the Nobel Committee arrived, and Luis went in to the lab to find that his office was festooned with messages and strewn with balloons labeled with the names of new elements and subatomic particles the scientists at the Rad Lab had discovered.

A press conference with the president and chancellor of the university gave him a chance to recognize colleagues whose work had contributed to the award. Luis's citation for the Nobel noted "his decisive contributions to elementary particle physics, in particular the discovery of a large number of resonance states, made possible through his development of the technique of using hydrogen bubble chamber and data analysis." In other words, he bombarded atoms with subatomic particles to see what

Luis Alvarez celebrates news of his Nobel Prize with colleagues.

the atoms would do and what else could be found in the nucleus, and he developed new methods for doing this work. He had worked with hundreds of scientists, fellow faculty, students, and staff members in pursuing these goals, and he offered them his thanks.

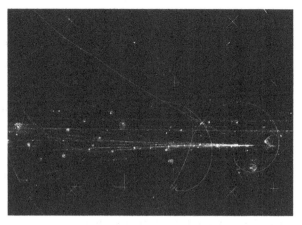

A high-speed image of the paths of subatomic particles in a hydrogen bubble chamber.

Over the next six months, Luis received over 750 cards, letters, and telegrams of congratulations. He responded to each one with a personal note. With his prize money, he invited and paid for eight colleagues, who had contributed to the research that won the award, to bring their wives and attend the Nobel ceremonies with him and Jan. Walter and Milly flew from Libya to Sweden to share in the joy.

Bubble chamber, Berkeley, 1954.

Over the next several years, Luis kept his commitment to maintaining his existing relationships and to creating an environment in which researchers could collaborate and discuss ideas. Every Monday night, he and Jan opened their home to colleagues for a rigorous, lively presentation and discussion of a particular topic in physics or another science—a process Luis had learned from his first mentor at Berkeley, the Nobel Prize-winning physicist Ernest Lawrence. "Since most physicists don't really understand a subject until they've discussed it with their colleagues, talking physics is essential to doing physics," Luis said. "So we gathered weekly to discuss the nuclear-physics literature." Having time to think about ideas and understand how they related to each other not only helped the researchers, but also led to the advancement and expansion of the whole science of physics.

> "Talking physics is essential to doing physics."
> —Luis Walter Alvarez

Andy Buffington was a young postdoctoral researcher—a scientist who has just completed a PhD—who worked in the A Group. Buffington said that the Monday-night discussions could become quite heated. "If Luis thought he was getting nonsense, he would keep at it rather acidly, and not let up," Buffington said. "It was a combination of friendly and hostile questioning." If a scientist was unable to understand a problem, or design an experiment, colleagues might help them think through ideas. Collaborative discussions like this helped them work as a team to puzzle out solutions.

Scientific discussions can sometimes sound like arguments, when one scientist is challenging another to defend their ideas and conclusions. Like the peer-review process, however, scientific discussions help researchers grow. Science is about asking questions and finding evidence. Discussions help make sure the evidence and data answer questions accurately.

A FAMILY AFFAIR

While Walter and Luis kept in touch, and Walter and Milly had attended the Nobel ceremony with Luis and Jan, the men knew little about each other's scientific work throughout the 1960s. Then, in 1971, Walter and Milly returned to the United States from their research in Italy and Libya, and Walter began work at the Lamont lab in New York. Luis came to visit. While he had not been interested in geology earlier in his career, Luis's interests had broadened as he made use of his own father's advice to let his mind wander and see connections among subjects. In recent years, he had used cosmic rays to try to find hidden chambers inside a sealed Egyptian pyramid—unsuccessfully, as the pyramid was made of solid rock. Now, as he learned more about Walter's work as a geologist, Luis became fascinated by the new discoveries in plate tectonics. Father and son vowed to find a way to collaborate on a project in the future.

Throughout the early 1970s, Walter continued to travel to Italy, collecting rocks from different Scaglia rossa outcroppings in Gubbio. No matter where he looked in the Apennine mountain range, there it was, that thin layer of gray clay between the Cretaceous and Tertiary layers. His mind remained caught by the questions it presented: Where did the clay come from? Why was it there? How long did it take to form? And, most important, did the clay layer have anything to do with the K-T mass extinction?

Walter puzzled especially over the length of time it had taken the clay layer to form. Clay was composed of extremely fine particles. Some of those particles could have settled down from dust in the air and atmosphere. Much of it would have washed down from high mountain lakes, settling as a layer in bays and in the mouths of rivers. The Law of Superposition states that the thickness of any rock layer is an indication of how long it took to form: the thicker the rock, the longer the formation process. This clay layer was thin, which meant it must have been deposited in a short period of time—unless there was an **unconformity**, a gap in the geological record, when layers of sediment are washed or torn away and then covered again by new sediment.

Walter talked with his father frequently about the puzzle of the clay. Could physics help them understand how long it

Missing Pieces

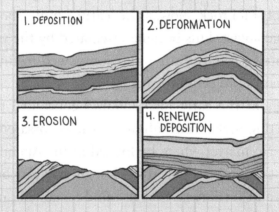

Sedimentary rock is created in layers as one layer of sediment piles on top of another. The oldest layer is on the bottom, while the newest layer is on top. But natural forces sometimes change these layers in unexpected ways. **Erosion** caused by wind, water, or ice can wear rock layers away. Earthquakes can tear off strata. Then, when a new layer of sediment covers the rocks, there is a gap in the geologic record. The place where the layers come together with missing pieces—the new boundary—is called an unconformity. The span of time missing in the stratigraphic record is called a **hiatus**.

An example of an unconformity from the stratigraphic record in Utah.

took the clay layer to form? They tossed around ideas, just as Luis had always done with his students. Then Luis went to the library.

"Luis was a 'want to know' person," Andy Buffington recalled years later. "He was skilled at using libraries to ferret out information." Even after getting his PhD, Luis had seen there were missing parts to his education, topics he needed to catch up on. "The only way I could learn nuclear physics," he concluded, "was to read *everything* that had been written on the subject."

After reviewing the literature on the clay layer, Luis had an idea for how to solve the puzzle. What if they used radiometric dating to measure how much time passed as the layer was deposited? They could do this by measuring how much of the **isotope** beryllium-10 was in the clay. Beryllium-10 (Be-10) had a half-life of 2.5 million years—just slow enough to be present in the clay as an indicator of how much time passed as the layer formed. Luis suggested Richard Muller, a Berkeley physicist, to do the measurements.

In 1976, Muller paid a visit to the Lamont lab to lecture about radiometric dating. He soon began to work with geologists to measure the ages of sedimentary rocks in the ocean floor and figure out how to use physics to solve other problems in earth science. He and Walter were in the midst of plans to measure the Be-10 in the clay when they learned some disappointing news: the half-life of Be-10 was actually much shorter than previously thought—only 1.5 million years. If Be-10 had ever been present in the clay at all, it was unlikely there would be a measurable amount left 66 million years later. They would need to find a different element to test.

In 1977, after ten years of international travel and Walter's postdoctoral work at Lamont, Walter and Milly were ready to find a permanent home. When a job opened up at UC Berkeley, Walter applied and, to his great delight, was hired as a professor. He and Milly returned to his hometown to live, work, and hopefully continue to collaborate with Luis.

In Berkeley, Walter and Luis had more time to work together. While they both wanted to find out how long it took for the clay layer to form, they realized they also needed to discover *how* the clay layer formed in the first place. They knew that the surrounding limestone was made from the shells of foraminifera and other microscopic creatures that drifted to the bottom of the ocean at a steady pace and were compressed into rock. They imagined two ways that the clay could have been deposited on top of that Cretaceous limestone. In one scenario, the limestone could have formed at an ordinary rate, but then suddenly a huge amount of clay washed into the ocean all at once, leaving that one-centimeter layer on top of the limestone. In another possible scenario, clay washed into the ocean regularly, mixing with limestone. Then, for a period of time, no shells or limestone were deposited at all, leaving only the layer of clay.

Luis and Walter needed a way to test which scenario might have been true. They needed to find a substance in the clay that was deposited on Earth at a steady, measurable rate. Something, perhaps, that came from outer space.

Something like star dust.

MEASURING STAR DUST

The Alvarezes knew that in addition to cosmic rays, a steady stream of **cosmic dust** rains down on the earth all the time. In fact, an estimated 10 million kilograms of cosmic debris—mostly dust—falls to Earth each year. In that dust is an element very rarely found in the earth's crust: iridium.

Iridium is an extremely heavy metal. When the earth formed, most of its iridium sank deep into the planet's core, leaving little on the surface. In space dust, however, iridium is abundant. If the Alvarezes could measure how much iridium was in the clay layer, it might tell them how long it had taken to accumulate the dust. Clay that had been deposited steadily over thousands of years would have a traceable amount of iridium. A thin layer of clay that was deposited quickly, over a decade or less, would have very little iridium.

To measure the samples, the Alvarezes turned to Frank Asaro, a nuclear chemist. Measuring star dust in the layer of clay involved working with tiny fractions called "parts per billion," or ppb. In a sample of clay divided into a billion parts, how many parts would be iridium? Since cosmic dust settles on the earth at a rate of ten millimeters (one centimeter) every thousand years, a centimeter of clay containing a measurable amount of iridium would take at least a thousand years to form. Luis and Walter theorized that if the dust settled slowly over a thousand years, they might find 0.004 ppb to 0.01 ppb of iridium in the clay.

To test this hypothesis, Asaro used a system called "neutron activation." First, he put the clay sample into a chamber with a "reactor," a radioactive substance that can cause a sample to react, or give off subatomic particles. As the reactor fired neutrons at the clay sample, Asaro watched what happened. This bombardment made some atoms

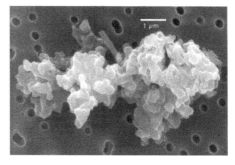

Particles of cosmic dust.

in the sample decay and release a **gamma ray** as a single photon, a particle of light. Asaro had discovered that the gamma rays from different elements gave off differing amounts of energy. By looking at the specific energy signals coming from the gamma rays, he could tell how many parts per billion of each element were in the clay.

Luis Alvarez and Frank Asaro in front of a mass spectrometer.

It is very easy to make mistakes in a process this complex. It took extreme precision to get accurate results. When Luis and Walter asked Asaro for help in October 1977, he had just finished doing an analysis of some ancient pottery pieces, trying to determine their age. He was about to do the same kind of analysis on soil and volcanic ash to find out when volcanoes had erupted in ancient times. He told Walter and Luis he didn't have time to add another project to his roster and still be as accurate as he needed to be.

Asaro also had reservations about finding iridium in the limestone above and below the clay. Iridium was very rare. He had not found any detectable iridium in the many soil and rock samples he'd studied for other projects, nor in the pottery samples he'd recently analyzed. What made the Alvarezes think this clay and the limestone around it would be any different?

Walter and Luis explained that they planned to dissolve the limestone in acid. It was possible that this would release iridium, making it easier to measure. They also explained that even if Asaro didn't detect any iridium, there were twenty-five other elements he might find that would ultimately move the project forward. Asaro thought the project sounded intriguing, and he was eager to work with Luis. If they were willing to wait until he completed some prior projects, he agreed to test their samples for iridium.

Walter brought twelve samples of the Scaglia rossa limestone, from the clay layer as well as above and below the clay, to Asaro's lab and dropped them off. Each one was carefully labeled with exactly where the sample had been found, its location in the strata, and when he had collected it. Then he waited.

And waited.

And waited.

Months went by. Frank Asaro was frustrated too. A broken gamma ray detector had created a backlog of three hundred other samples that needed to be tested before he could get to the Alvarezes' work.

Finally, in April 1978, Asaro put a small sample of the Scaglia rossa clay into the nuclear reactor for eight hours of irradiation. He detected no gamma rays from iridium. He tried again, using a larger sample of clay from the upper portion of the K-T boundary, with most of the limestone dissolved away in acid. It was irradiated in a stronger reactor for 224 minutes.

In June 1978, Asaro called the Alvarezes into his office. He had completed the analysis, but the results were not what they anticipated. Where they were looking for 0.004 ppb to 0.01 ppb of iridium, Asaro had found 3 ppb of iridium in his second analysis—a thousand times what they expected. Later, they would discover that iridium can be lost in the acid used to dissolve the limestone, so the real reading was closer to 9 ppb.

Walter and Luis were astonished. What on earth could have caused such a large amount of iridium to appear in the clay?

What on earth? Nothing on Earth.

The source must have come from outer space.

IT CAME FROM OUTER SPACE

When the Apollo 11 mission landed on the moon in 1969, astronauts saw that the desolate lunar surface was pocked with craters of all sizes, from gargantuan gorges to pits as small as the head of a pin.

The astronauts collected moon rocks and lunar dust. Back on Earth, scientists discovered the dust contained pieces of broken rock and mysterious tiny spheres of natural glass, called **tektites**. Under a microscope, they could see that each tiny tektite was pitted with craters. This was a clue that the moon was once bombarded by interstellar rocks and dust.

In fact, Mars and other planets and **asteroids** were all dimpled with impact sites—holes caused by collisions with massive objects from space. **Impact craters** were clearly commonplace in the solar system.

Some scientists wondered if Earth was once bombarded too.

Evidence from space changed a lot of ideas about how Earth formed. But one idea seemed stuck in time: uniformitarianism. For 40 years, J. Harlen Bretz challenged geologists' ideas about how Earth changed over time.

1920s

When Bretz, a geology teacher with a PhD, looked at the scarred landscape of eastern Washington, he didn't believe it could have formed slowly.

To him, the channeled scablands looked like the site of an ancient catastrophe.

He began to think that sometimes Earth *could* change in an instant.

In 1923, he proposed that a catastrophic deluge, which he called the Great Spokane Flood, had poured through the region during an ice age and carved the land.

Since Bretz couldn't explain what caused such a flood, his idea was ridiculed.

"Ideas without precedent are generally looked upon with disfavor and men are shocked if their conceptions of an orderly world are challenged."
—J. Harlen Bretz

Nonsense!

Boo!

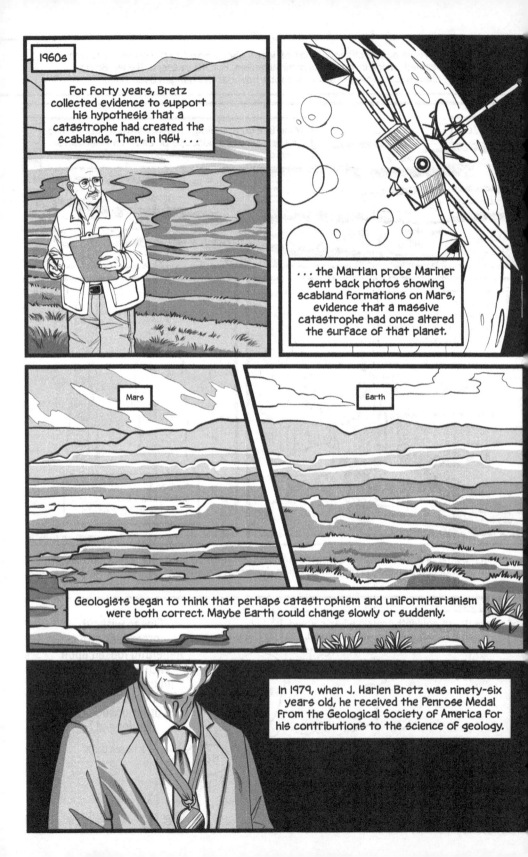

CHAPTER FOUR
HIGH-IMPACT EVIDENCE

Team Alvarez needed to see if iridium was present in high concentrations in the K-T boundary worldwide. Following Luis's lead, Walter went to the library to search for information about other K-T boundary sites. At last, he found Stevns Klint, a site in Denmark just south of Copenhagen, where clay was sandwiched between chalk and limestone in the exposed cliffs.

In 1977, Walter traveled to Gubbio to collect more rock samples. Then he flew on to Copenhagen to meet with Danish scientists. He was met by Soren Gregersen, a seismologist who studied earthquakes, and Inger Bang, a micropaleontologist who studied microscopic fossils. The three scientists drove to Stevns Klint, a seaside community on an island an hour from the airport. They parked by the side of the road on the broad, flat top of a grassy cliff. Taking field tools for collecting rocks, they scrabbled down the cliff to the edge of the dark blue water. Small, smooth gray stones clacked under their feet as they clambered to the shoreline and looked up at the rock above their heads. The cliff was rich with beige limestone and layers of softer, white chalk, both of which were made of the microscopic fossil shells that fell to the ancient ocean floor.

Stevns Klint, Denmark.

Walter's nose found the clay layer as soon as his eyes did. A sulfurous smell emanated from a thin layer of black clay in the cliff. "Fish clay," he called it, as lifeless as a cemetery. It was clear to him that something had happened to turn the beautiful blue ocean, represented by the white chalk and limestone, into a deathly black bed of clay, empty of fossils. Something catastrophic.

The three scientists collected samples of the clay and the limestone and chalk above and below it. They labeled the samples with their exact locations, wrapped them up, and tucked them into a sack for Walter to haul back to Berkeley. Then they climbed up the cliff to the car. A clean, clear breeze from the ocean carried away the sulfurous smell of the rocks as they sped to Copenhagen.

At Berkeley, Frank Asaro irradiated the clay from Stevns Klint, looked at the gamma ray data, and analyzed the results. It also had a high concentration of iridium. Now, with two European K-T outcroppings confirming the same unexpected results, Luis and Walter let their minds turn to a possible cause. What force could have left parts of the earth covered in iridium?

Luis went back to the library and began to read. As always, he read everything he could get his hands on, but this time he focused on space—specifically, a supernova.

SUPERNOVA

A supernova is an exploding star. Stars, such as our sun, shine because the nuclei of hydrogen atoms fuse together to become helium, releasing energy. When this hydrogen atom fuel in a star runs out, the star dies by collapsing in on itself. As it collapses, elements ricochet into each other as the outer layers of the star bounce around and careen off the iron core, exploding outward in a cascade of cosmic radiation and elemental dust. Life on any planet in the vicinity of a supernova explosion would likely be wiped out.

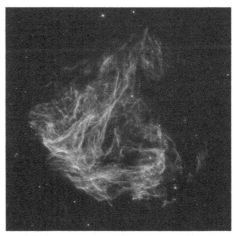

Radiation from a supernova would wipe out life on any planet nearby.

In 1971, two scientists, Dale Russell and Wallace Tucker, hypothesized that a supernova could have caused the K-T extinction. They explained that when a supernova explodes, it releases large pulses of cosmic rays. This radiation could push the temperature near the star to over a million degrees. If a supernova had exploded near Earth 66 million years ago, it would have baked the upper atmosphere—the ozone layer and the ionosphere, between 20 and 100 kilometers above—and drastically changed atmospheric circulation. This in turn could have created supersized storms and an ice age that would have dropped the earth's temperature, resulting in mass extinctions.

Luis read about the possible effects of a supernova to life on Earth. Then he called a colleague to throw around ideas. Like the weekly discussions Luis held in his living room, casual conversations are a way scientists test out ideas, get feedback, and brainstorm possibilities.

Andy Buffington was home in his living room when Luis called. Luis explained their iridium findings, and they pondered what could have caused this anomaly. Luis wondered: Could iridium be evidence of a supernova? Did a supernova wipe out the dinosaurs?

Buffington shook his head. "You're barking up the wrong tree," he said. He crossed the room and pried a beige volume of the encyclopedia off his shelf. He thumbed through pages until he came to the entry he wanted: "E for Extinction."

The problem with the supernova idea, he told Luis, was that it only worked once. He ran his finger down the text of the book and read it over the phone. The K-T boundary extinction, while the most famous, was only one of five mass extinctions on Earth. Wouldn't it make more sense to find a cause that related to more than one extinction? It seemed unlikely that all five could have been caused by separate events, especially something as rare as a nearby supernova, which happened about once every billion years.

Luis pointed out that some people had noticed regular intervals between mass extinctions. Maybe these extinctions happened on a regular basis. He tossed out another theory: Perhaps the sun had a twin star, a "nemesis" star, with which it orbited around a central point of dense gravity every 23 million years. Perhaps it came close enough to deposit iridium on Earth every 23 million years. But the odds of that happening were, according to Luis's calculations, "one in a billion in a hundred million years."

After he got off the phone with Andy, Luis knew he still wanted to investigate whether a supernova could have caused the extinction. To do that, he needed to find another element that would be present in the clay if a supernova had irradiated Earth. Luis decided to look for plutonium-244, a common element emitted from a supernova. With a half-life of 83 million years, plutonium-244 would still be detectable in the Gubbio clay if a supernova had caused the extinction.

To test the sample, the team turned to Helen V. Michel, the nuclear chemist in charge of the neutron activation team at the Lawrence Berkeley Laboratory. She would work with Frank Asaro to bombard the clay sample with radiation to try to detect the plutonium-244. There was one challenge: because the chemical could quickly decay to the point that it became undetectable, there would be no time for breaks.

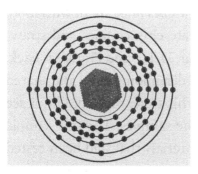

A model of plutonium-244.

For thirty-six hours straight, Helen and Frank bombarded the clay samples from both Gubbio and Denmark, observing and calculating the resulting gamma rays. Walter, Luis, and Milly Alvarez plied the pair with coffee, pizza, doughnuts, and chili so they could run continuous tests. At last, in the chill Berkeley dawn, they delivered their results: the samples tested positive for plutonium-244, proof positive that a supernova had caused the extinction of the dinosaurs.

Luis was thrilled. At last, here was the answer to a question the world had been asking for over a century. How did the dinosaurs die out? A supernova! He was ready to shout the news from the rooftops. But Walter wasn't so sure. He and Frank went to the deputy director of the Lawrence Berkeley Laboratory, Earl Hyde, to ask for advice. Hyde was a nuclear chemist like Frank. Did he think they should release the information right away?

Hyde asked them to explain their experimental procedure. Exactly how did they get their information? What steps did they take? They knew as well as he did that one of the most important rules of science was repetition. For an experiment to be considered valid and accurate, it must be repeatable.

"Do it all over again," Hyde told Walter. "Repeat every single step from the very beginning, on a fresh sample, to be absolutely sure there really is plutonium-244 in that clay."

Back at the lab, Helen and Frank returned to work. Once again, they bombarded the carefully labeled samples of clay with radiation, observed and counted gamma ray emissions, and analyzed the results. Once again, the Alvarezes plied them with food, including Milly's homemade chocolate-chip cookies and strawberry ice cream. Thirty-six hours later, Helen and Frank sat back, their faces slack with exhaustion and disappointment, and shared their results with Walter and Luis. This time, they could not see a single trace of plutonium-244 in the samples. The first sample had likely been contaminated by materials that had been tested on the equipment earlier. The supernova answer was out.

 "Do it all over again. Repeat every single step from the very beginning, on a fresh sample to be absolutely sure." —Earl Hyde

A NEW IDEA EVERY WEEK

Now that they had discarded the supernova idea, Walter and Luis began looking for other possible causes for the K-T extinction. For months, they discussed ideas with friends, puzzled, and dreamed, searching for a connection between the iridium and the extinctions.

Perhaps a series of tidal waves had wiped out the animals. Luis's friend UC Berkeley astronomy professor Chris McKee extended the idea: What if an asteroid bearing iridium crashed into the ocean, and the resulting tsunami wiped out animal life? Luis argued that a tsunami would have impacted only one ocean. What kind of wave was big enough to wipe out over 75 percent of species on Earth? They dismissed that idea and moved on to the next.

Perhaps Earth had passed through giant nebular clouds full of hydrogen, which combined with oxygen to remove the oxygen store from the earth. If this were the case, however, plant life would have been just fine, and it would have eventually replenished Earth's oxygen. They threw out this idea too.

Maybe hydrogen and iridium had been knocked loose from Jupiter and had somehow ended up on Earth. Perhaps a super solar flare sent iridium particles streaming down to the planet. Frank Asaro later recalled that Luis came up with a new idea every week for six weeks, bouncing ideas off colleagues and tossing them out, one after another. They were looking for an event that potentially could have repeated in each of Earth's five major extinction events. But how often did solar flares ignite the sky? How often did Earth pass through nebular clouds?

At last, they returned to an earlier idea, an occurrence that was surprisingly common. Maybe they had dismissed it too quickly before. How often did meteors zoom through Earth's atmosphere?

How often did asteroids hit other planets or our moon? Pretty often, if the craters in the rest of the solar system were any indication. An extraterrestrial impact could have sprinkled the earth with interstellar iridium. But how could one impact in a single place wipe out life on the entire planet?

Asteroids, Meteors, and Comets, Oh My!

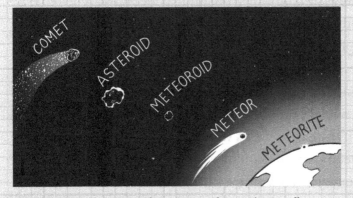

Comets, asteroids, meteoroids, meteors, and meteorites are all objects smaller than a planet moving through space.

If it looks like a shooting star, it's probably . . . any of the above. An asteroid is a large chunk of rock that orbits the sun. Meteoroids are rocks—from the size of a speck of dust to a chunk of a small asteroid—moving through space. When meteoroids or asteroids enter Earth's atmosphere and begin vaporizing like shooting stars, they are both called meteors. If a meteor actually reaches Earth as an intact rock, it is called a **meteorite**.

The difference between asteroids and meteoroids is rather inexact, though it has to do with size. Asteroids can vary from the size of a small car to the size of a large city. (The largest asteroid, called Ceres, at almost 966 kilometers [almost 600 miles] in diameter, is actually considered by some to be a dwarf planet.) Meteoroids are typically much smaller.

In contrast to these rocky space objects, comets are balls of ice, gas, and dust that orbit the sun, streaming a trail of glowing, gassy debris behind them. They sometimes collide with planets, such as when scientists observed the comet Shoemaker-Levy 9 smash into Jupiter in 1994.

Meteorites don't burn up completely in the atmosphere.

ALL TOGETHER NOW

Luis closed his eyes and stretched his thoughts back to the broad body of knowledge he had been amassing throughout his life. He let his mind wander and wonder, as his father had told him to do, collecting and connecting information. He thought about witnessing the mushroom cloud of a nuclear explosion rising to 40,000 feet over New Mexico. He thought about space travel and the craters the astronauts had seen on the moon. He thought about the Ring of Fire in the Pacific and winters in Minnesota.

Luis considered the kinds of impacts that asteroids and comets had already made on Earth. He read about a comet that had exploded over Siberia, Russia, in 1908. It partially vaporized and then fragmented into chunks of rock and dust, engulfing large swaths of forest in flames.

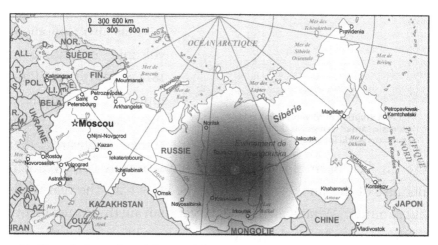

The Tunguska impact in 1908 destroyed almost 2,072 square kilometers (over 800 square miles) of forest.

What if a comet or asteroid bigger than the one that hit Siberia plummeted down to Earth? How large would its path of destruction be? How many kilometers of forest would burn if it were five times, ten times, a hundred times larger than the Siberian object?

Luis also thought back to the days when his grandfather was living on the volcanic islands of Hawaii. He pulled out a book his father had given him about the 1883 eruption of Mount Krakatoa in Indonesia. The book had been published in 1888, so it offered a ringside view of the catastrophic event, complete with interviews with observers, scientific and mathematical calculations of the explosion, and a historical picture of how it affected the planet.

Luis also read about volcanic eruptions in 1814 and 1815 in the Philippines and Indonesia that had sent so much smoke skyward that the sun had been blocked, causing unseasonably cold weather. Because of the smoke, 1816 came to be known as "the year without summer." What if an impact, larger than anything seen before or since, caused a world without summer, a life without the sun, a planet without life?

Luis began to formulate a hypothesis. On July 14, 1979, he wrote a letter to Walter, who was doing fieldwork in Italy, and to Helen and Frank. What if a comet or asteroid grazed Earth, breaking

The 1883 eruption of Mount Krakatoa in Indonesia caused changes in climate patterns around the world for several years.

1816 was called the year without summer because smoke in the air from volcanic eruptions blocked sunlight, causing many parts of the earth to cool.

up in the atmosphere and coating Earth in dense smoke? "The smoke would obscure the sun, turning off photosynthesis, and the animals would die of starvation."

Walter, Helen, and Frank were critical. Grazing events, in which an asteroid brushes up against Earth's atmosphere, are extremely rare. Luis tried again. What if a pulverized chunk of space rock hit Earth directly, so hard that the object vaporized, sending so much smoke, ash, and cosmic dust into the air that it blocked the sun for months? What if the whole world was affected? What if the impact of an asteroid or meteor *could* cause a worldwide extinction event?

This time the other three scientists thought Luis was getting closer to the truth.

Luis dove into the mathematics behind an impact. He read the work of Sir George Stokes, a nineteenth-century scientist who studied the speed at which particles fall through the air and how long they stay suspended, floating in the atmosphere. After the eruption of Mount Krakatoa, Stokes was able to calculate the size of the ash and dust particles left in the air by measuring the rings of light that shone around the moon. He predicted that the particles would stay in the atmosphere for two years—even in Europe, on the other side of the globe from Indonesia. He was right. The red sunsets, caused by the particles, continued for twenty-four months after the eruption.

Smoke Gets in Your Eyes

In 1851, British physicist and mathematician George Stokes published a paper describing the velocity (or speed and direction) of small spheres moving through a thick, semi-liquid material. Small objects, such as particles of dust, fall through air and water at a predictable rate that can be calculated. The smaller the sphere, and the thicker the material it is falling through, and the more turbulent—windy or wavy— the material is, the longer it will take to fall. This is known as **Stokes's Law**. Walter, Luis, and other scientists would use it to calculate how quickly dust would fall to Earth or how fast sediment would sink to the bottom of the ocean following an impact.

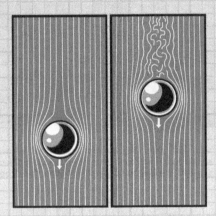

Stokes's Law helped Walter and Luis calculate how long it would take for dust from an impact to fall to Earth.

By calculating the amount of iridium found all over the earth, and what was known about how much iridium is typically found in meteors, Luis figured that an object ten kilometers in diameter hit the earth approximately 66 million years ago. The impact of that object could have thrown enough dust into the atmosphere to darken the sky for years. This would have killed plants by preventing sunlight from reaching them, which would eventually kill all plant- and then meat-eating animals that weighed above fifty pounds all over the earth. When he was sure of his calculations, Luis called Walter in Italy. "We've got the answer!" he said. He insisted Walter present the findings at a meeting in Copenhagen.

While Walter and the team were excited by Luis's hypothesis, they were also cautious. Walter wondered how scientists who had always believed in a uniformitarian model of how Earth changes would feel about a catastrophe. He was sure there would be great resistance. Nevertheless, in September 1979 Walter presented the team's preliminary findings—without mentioning the impact hypothesis—at a meeting about the K-T boundary extinction that took place in Copenhagen, Denmark. He described the iridium, the lack of plutonium-244, and the locations they had examined so far. Then he waited for the debate to begin.

Immediately, scientists began to approach him to share their own discoveries. The first was Jan Smit, a Dutch scientist who had discovered a clay layer in Caravaca, Spain, around the same time that Walter and Luis were testing the Gubbio clay. In reading the results of the Alvarezes' study, Smit noted that his sample also contained unusually large amounts of iridium. When another scientist sent evidence of high iridium levels in rocks from New Zealand, the Berkeley group knew it was time to explain their impact hypothesis.

In June 1980, Walter Alvarez, Luis Alvarez, Helen V. Michel, and Frank Asaro published a peer-reviewed paper in *Science* magazine, "Extraterrestrial Cause for the Cretaceous-Tertiary Extinction," theorizing that an extraterrestrial impact somewhere on Earth caused the K-T extinction and the end of the dinosaurs.

Helen V. Michel, Frank Asaro, Walter Alvarez, Luis Alvarez, 1969.

Like all scientific studies, the article served as a call to action for other scientists. Could they find evidence of iridium in K-T boundary rocks from around the world? Could they find other clues, such as an impact site? Or would they bring evidence that would disprove the hypothesis altogether?

NASTIEST FEUD IN SCIENCE

Immediately, other scientists began lining up for or against the impact hypothesis. Many paleontologists noted that while a meteor may have struck with worldwide consequences, the evidence from dinosaur bones suggested the impact happened long after the extinction of the dinosaurs. On his dinosaur digs, Bill Clemens, a colleague of Walter's at UC Berkeley, discovered an iridium layer three meters above the last dinosaur bones. Other scientists who had spent lifetimes researching volcanic flows believed volcanoes had wiped out the dinosaurs.

In those initial years, debates were heated, and not always polite, as scientists across different disciplines questioned the methods and findings of those in other fields. Luis Alvarez, passionately attached to his impact hypothesis, even launched insults at other scientists, comparing paleontologists to glorified "stamp collectors." Paleontologists who supported the volcanism theory, in turn, questioned why a physicist thought he knew more than dinosaur scientists who had spent a lifetime studying extinctions. The debate was dubbed the "nastiest feud in science."

Many scientists, however, were eager to find evidence that an impact had been the culprit in the dinosaur extinction. They reexamined and tested boundary rock samples they had collected from all around the world. In addition to Smit and the New Zealand scientist, researchers reported high levels of iridium at the K-T boundary in rocks from the Raton Basin of Colorado and the bottom of the Pacific Ocean.

Then a group in New Mexico discovered the same iridium-rich clay layer in rocks that came from a coal swamp, not the seafloor. This was important, because some critics claimed that iridium could have floated freely in a prehistoric ocean and not come from outer space.

Rock from the Raton formation in Colorado contained iridium.

The more K-T boundary sites contained iridium, the less likely it was that the impact hypothesis could be proven incorrect, because only a huge impact could have spread so much iridium over the entire planet. But Walter and the other scientists needed more than iridium to support their hypothesis. They needed to find ground zero. They needed an earthen scar, a gash so large that no one could doubt an impact had happened at the K-T boundary and that it wiped out three-quarters of the life on Earth. They needed to find the crater.

Coast Guard! Mayday, mayday. This is the *Edrie!* Come in. Mayday, mayday!

The Ulriches were two of the four known surviving witnesses of one of the largest tsunamis in recorded human history. The tsunami was not the first big wave to hit Lituya Bay, but it was the first anyone is known to have seen.

Tsunamis happen when an earthquake, a volcano, or an underwater landslide sends energy through the water, causing a giant wave to form.

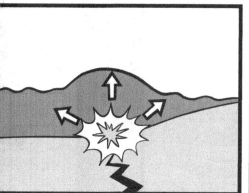
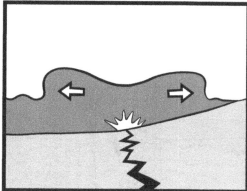

Unlike regular waves, which crash and recede every few minutes, tsunamis flow like an unstoppable moving wall of water for thousands of kilometers, pouring onto land for ten minutes at a time, and taking even longer to recede. As tsunamis flow, they move sediment and debris to new places.

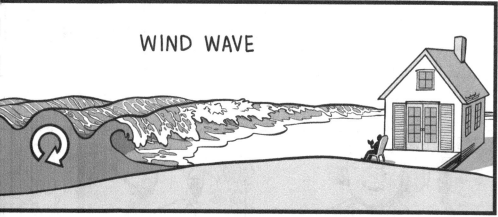

WIND WAVE

TSUNAMI WAVE

CHAPTER FIVE
CLOSING IN ON A KILLER: THE SEARCH FOR THE CRATER

Jody Bourgeois, PhD, professor of Geophysical Sciences at the University of Washington, felt the thick July mud suck at the soles of her old tennis shoes. June 1987 had been one of the wettest months in recent memory in south-central Texas, and Darting Minnow Creek was still running high. She scanned the brown water for poisonous water moccasin snakes and snapping turtles that other scientists had warned her about, but she wasn't too worried. She was a geologist, after all. She was used to trekking into unknown territory armed with only a rock hammer.

In the middle of the creek, she saw what she was looking for: a slick chunk of gray rock sandwiched in

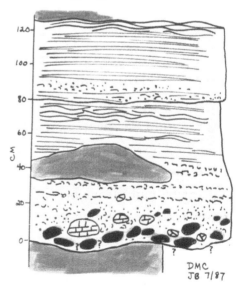

Bourgeois drew detailed pictures to capture what she saw at the Brazos River site.

layers of tan sandstone. To any other hiker, it might look like just a big rock in a stream. But Bourgeois knew it could be something else altogether: a sign of the K-T boundary, a bookmark stuck in time, an arrow pointing to the moment the world changed. She knew that this place, now far from the ocean, used to be part of the Western Interior Seaway, a long, shallow inland sea that covered much of western North America. And she suspected that something big had happened here long ago.

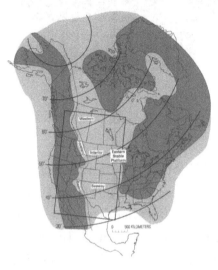

Map of the Western Interior Seaway, a shallow sea that divided what is now North America.

Bourgeois waded into the cool water and made her way through the shallow current to the smooth gray clast, or boulder, embedded in the sandstone. It looked as if some great force had ripped up a chunk of the muddy ocean bottom and mixed it with sand from an ancient beach. Could this formation have been caused by storm waves? Or could it be evidence of the megatsunami hypothesized by Walter and Luis Alvarez's recent asteroid impact idea? She had read the Alvarezes' paper, and had even heard Walter lecture on the subject, and she was curious. What kind of evidence would a giant tsunami leave behind? She would need to collect rock samples to know for sure.

As she examined the outcrop, a slimy brown leech squeezed itself out of a crack. Snapping turtles were one thing. Poisonous snakes—no problem. But blood-sucking leeches? Bourgeois gave a small shudder—then got back to work.

Bourgeois chipped away at different parts of the big mudstone with her rock hammer, placing small pieces into cloth sample bags. She labeled each bag with the date and information about the rock's location. In her science notebook, she drew pictures of the rock formations and added information like the weather and water conditions, rock types and wildlife. She took photographs too, but drawings were an even better way to record details of the site.

At the end of the day, Bourgeois hiked back to her truck in wet socks, as the mud had sucked the soles right off her shoes. She was pleased with the rewards of her labor—a backpack full of rocks and meticulous notes—and eager to get back to her lab to view the samples under a microscope. Soon she would have her answer.

TAKING SIDES

In the seven years since Walter, Luis, Helen, and Frank had published their paper hypothesizing that an extraterrestrial impact had caused the K-T extinction, scientists all over the world had taken on the challenge of finding a crater. They wanted to discover proof that an impact was connected to the extinctions—or else evidence that the impact and extinctions were completely unrelated. The scientific process was slow and painstaking. But Walter and Luis were in no hurry. They knew that good science—finding clues, measuring and analyzing data, collaborating with other researchers, and deciding how it was all connected—could take a lifetime.

Throughout the 1980s, Walter read every paper published on the subject. Some researchers didn't believe a meteor could have hit Earth at the time of the Cretaceous extinction. After all, where was the crater? The controversy took on an almost religious fervor with geologists searching for clues and lining up data either for or against the impact hypothesis, writing papers and presenting them at conferences. Egos flared as researchers challenged each other's ideas, research methods, and evidence about what caused the K-T extinctions. But challenging hypotheses is how science works, and in the end, scientists celebrate new and better explanations, even when data proves their own ideas wrong.

Some researchers questioned whether the iridium came from outer space. Since most known K-T boundary sites had once been in the ocean, these researchers claimed that the iridium in the

Meeting of the Minds: Scientific Conferences

Besides writing papers and talking informally to colleagues, scientists also present their findings at conferences where they discuss new ideas, introduce new evidence, and analyze what that evidence really means. The problem of what caused the Cretaceous extinction drew in experts from across the sciences, and it was important to meet to share ideas.

In 1981, 110 scientists from around the world were invited to Snowbird, Utah, for a conference titled Large Body Impacts and Terrestrial Evolution. Organized by scientists at the Lunar and Planetary Institute (LPI) and the National Academy of Sciences, the conference's intention was to allow scientists of many disciplines to discuss what happened at the end of the Cretaceous, and to teach each other the unique language of their own areas of science. Fifty-five scientists spoke and shared their research. Physicists discussed the energy of impacts; atmospheric researchers lectured on the chemistry of air; paleoecologists shared information about ancient environmental conditions; and statisticians looked at the mathematics of how to tell what happened when you have incomplete data. The meeting opened up conversations, essential discussions that have become more commonplace as scientists study worldwide problems together. Scientists met for this conference three more times between 1981 and 2000. Forty years later, paleontologist Jan Smit remembered that first Snowbird conference as "one of the best conferences I ever attended. Such enthusiasm!" with lots of time to discuss his questions with colleagues from around the world.

Scientists from different disciplines met at the Snowbird, Utah, conference to discuss the causes of mass extinctions.

clay must have come from ocean water and could not be extraterrestrial. Then scientists discovered iridium-rich K-T boundary sites throughout the Rocky Mountains, from New Mexico up to Alberta and Saskatchewan in Canada. Because these mountains had never been underwater, the discoveries confirmed that the iridium could only have come from sediment in the air drifting down to Earth—most likely from an asteroid or meteor, though possibly from a volcano.

Other scientists, such as geophysicist Charles Officer, PhD, of Dartmouth College, pointed to evidence that the dinosaurs were already dying out at the time the impact appeared to have happened. "The dinosaur extinction . . . was a gradual process that began seven million years before the end of the Cretaceous, and accelerated rapidly in its last 300,000 years" before the impact, Officer wrote in his book *The Great Dinosaur Extinction Controversy*, in which he reported on research by other scientists. Although he had not done his own paleontological field research, his readings indicated that many marine and aerial reptiles were dying out or already gone by the end of the Cretaceous. While other scientists disagreed with this analysis because dinosaur fossils are very rare to begin with, Officer's readings led him to believe that an impact could only have been a "capstone event," an event ending a process that was already happening on land, in the air, and in the oceans, and not the main cause of extinction.

Over the years, one paleontologist in particular, Gerta Keller, PhD, of Princeton University, looked for evidence that volcanoes played the main role in a Cretaceous extinction that started long before the K-T boundary event. Keller studied the role of traps, a type of volcanic flow that poured out floods of lava for thousands or even hundreds of thousands of years. Keller estimated that volcanoes called the Deccan Traps in India released lava for

350,000 years before the Cretaceous impact and 25,000 years afterward, eventually covering half a million square kilometers in volcanic rock up to 2,000 meters deep.

Keller hypothesized that pulses of flowing lava could have caused the Cretaceous extinctions by spewing so many greenhouse gases into the atmosphere with every pulse of lava that the climate changed worldwide. Scientists gravitated to one hypothesis or the other—impact or volcanism—as the search for evidence continued.

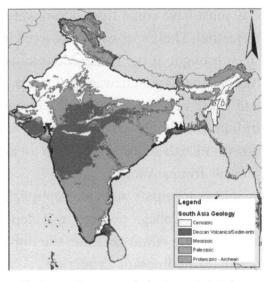

The Deccan Traps covered what is now India in lava up to two kilometers deep.

The Western Ghats hills at Matheran in Maharashtra, India.

FEATHERY FERNS OF CRYSTAL

While scientists such as Keller focused on providing alternative explanations for the extinction, Walter and other supporters of the "Alvarez hypothesis" continued to search for evidence of a crater. Finding the site of an impact would go a long way toward proving that their hypothesis was correct.

To find the crater, Walter took a new look at the clay from the K-T boundary to see what it might contain besides iridium. After all, if the impact was powerful enough to vaporize an asteroid, it must also have vaporized the earth's crust where it fell. If so, the K-T boundary clay should contain traces of that crust. If the impact happened on land, the clay would contain silicon, potassium, and sodium—the elements in quartz and feldspar that make up much of the **continental crust**. If the impact happened in the ocean, the clay would contain calcium and magnesium, the ingredients of the mineral basalt, which constitute the ocean's crust. Scientists began analyzing the clay layer to determine its chemical composition.

Walter's colleague Jan Smit, who, by Walter's estimation, had studied more K-T boundary sites than anyone else in the world, reexamined clay deposits from a site in Spain and found something unexpected: curious tiny white spheres, each the size of a grain of sand, embedded throughout the clay. He called them **spherules**. He cut the spherules in half, glued them onto glass slides, and then ground them down further to make tiny slices of each one, which he could look at under a microscope.

What he saw was surprising. When melted minerals crystallize slowly, they grow in distinct geometric patterns with sharp, straight lines. When they harden instantly, they form smooth, featureless

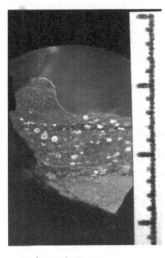

Under a microscope, spherules embedded in sediment show evidence of whether they came from continental or oceanic crust, offering a clue about where a meteor might have hit.

glass. The crystals in these spherules looked feathery, like microscopic ferns of frost forming across a window in winter. This pattern of crystallization could only have happened if the minerals had cooled neither quickly nor slowly, but at a moderate pace.

Smit discovered the spherules were made of sanidine, a kind of potassium feldspar, which would indicate they came from continental crust. When Smit examined the surrounding clay further, however, he found that it contained not only chemicals from the meteor and the rock that it struck, but also chemicals that had seeped into the site over millions of years. This caused the spherules to retain the feathery shape of the original cooling crystals even when new minerals seeped into the rock, in much the same way that silica seeps into wood and turns it into wood-shaped rock called petrified wood. As a result, Smit concluded that the original rock and the spherules actually consisted of **oceanic crust** infused with continental minerals, and that the impact had taken place somewhere in the ocean. Other geologists studied spherules in the boundary clay from all over the world and confirmed Smit's discovery.

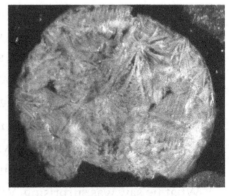

When Jan Smit looked at the spherules under a microscope, he saw feathery crystals, indicating the minerals had cooled at a moderate rate.

SHOCKING EVIDENCE

Even with this evidence that the impact likely happened in the ocean, many scientists continued to look for evidence of an impact site on land. In 1987, a team of scientists from the US Geological Survey discovered tiny grains of quartz with an unusual quality: they were cracked and split in distinct bands. Quartz with this lined structure, called shocked quartz, forms only in reaction to an impact. When a rock's crystalline structure is compressed and released by an earthquake, the rock absorbs the shock and returns to its original shape with no lasting damage. Rocks found both on the moon and at nuclear impact sites on Earth, however, carry permanent fine-lined cracks—the result of being smashed and crushed by massive shock waves. Shocked quartz on land was evidence that the impact may have happened in continental crust.

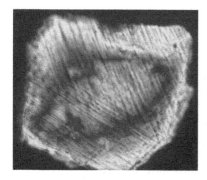 

Pictures of quartz shocked by an impact reveal clear bands of cracks not evident in quartz shocked by volcanic eruptions. Close-up on the right was taken with a scanning electron microscope.

Some scientists argued that a massive volcano could have produced shocked quartz as well. For months, Bruce Bohor, a USGS geologist, carefully examined the evidence, looking for differences in quartz that was known to be shocked by volcanoes versus quartz that was known to be shocked by an impact. He concluded that impact-shocked quartz produced multiple cracks in clear repeating bands, while quartz shocked by volcanic eruptions was erratic, with single lines of cracks without a regular, distinct signature. Shocked quartz was thus a new indicator of an extraterrestrial impact, and scientists began looking for it everywhere.

In 1987, Gene Shoemaker and Sandro Montanari found evidence of shocked quartz from a cluster of comet showers around 34 million years ago. They reasoned that the shocked quartz in the K-T boundary clay could be the result of several impacts, not just one asteroid. In some places in western North America, the K-T boundary minerals contained both spherules of glass and shocked quartz, providing evidence of both an oceanic and a continental impact. They wondered if one meteor had fallen on land, while another had fallen in the ocean. But if finding one impact crater was hard, finding two of the correct size and age seemed next to impossible.

CRATERS HERE, CRATERS THERE, CRATERS EVERYWHERE

One scientist had made it his business to record all of the verified impact craters in the world. Richard Grieve, a Canadian geologist, maintained a list of craters that he had carefully measured, dated, and recorded. Most craters were smaller than the size Luis estimated the K-T impact crater would have to be, based on the amount of iridium the meteor left behind. The Arizona meteor crater studied by Gene Shoemaker, for example, was just over a kilometer in diameter, that had been made by a meteor just fifty meters in diameter.

The Barringer Crater in Arizona was formed by an impact approximately 49,000 years ago. Today it is a tourist attraction and museum.

Many of the craters were in North America and Russia, areas with very old, exposed rock formations. Dating the craters was difficult and not always accurate, but Walter and his colleagues ran through the list, trying to identify which ones could have been formed by a K-T boundary impact. They found only three that were large enough to have caused a possible extinction event, and none of them appeared to have been created at the right time.

One crater stood out as a promising possibility. In Manson, Iowa, a crater thirty-five kilometers across had been covered over with glacial sediment and now lay under a farm. **Ejecta**—the materials thrown up in the air by an impact—found as far away as South Dakota indicated that a gigantic meteor struck the area. But *when* that happened was a question that could only be solved by careful data collection. Drilling down into the Iowa farmland, Gene Shoemaker and colleagues discovered a wealth of crushed rocks and shocked quartz—definite evidence of an impact. After dating the rocks, however, they discovered that the impact had occurred 75 million years ago, approximately 10 million years before the K-T extinction. A reexamination of the ejecta confirmed their suspicions: the ejecta had accumulated far below the K-T boundary. The Manson crater had nothing to do with the extinction of the dinosaurs.

Walter and his team refocused their efforts on finding the crater somewhere in the ocean. The only problem was that an impact in the ocean 66 million years ago would probably leave little evidence behind. Oceanic crust is constantly being renewed as new crust emerges from the mid-ocean ridge and old crust is recycled at plate boundaries. Twenty percent of the oceanic crust had been subducted since the K-T extinction. What if the impact site had been subducted and destroyed? Finding evidence in a remote part of the ocean would present another challenge. If the impact happened in deep ocean waters, how could they pinpoint the location?

Walter played possible scenarios over and over in his head. What would have happened if a meteor had fallen into the ocean? If a meteor ten kilometers in diameter smacked into the ocean at approximately 72,000 kilometers per hour, how big a tsunami would it cause? What kind of evidence would remain?

SIGNS IN THE ROCKS

Walter continued to collaborate with scientists all over the world in the search for the crater. Paleontologist Thor Hansen was researching fossil evidence at the K-T boundary in Texas when he noticed that rocks in the Brazos River area were very different from those he had seen before at other sites. Instead of clay, the boundary layer was made of sandstone and conglomerate—cobbles of stone cemented together with sediment—with fine-grained mudstone above and below. After sending the mudstone from the top to the Berkeley lab for testing, Hansen learned it was full of iridium. He invited other scientists to come to the river to research. After visiting the site and noticing evidence of churned-up sediment, Jan Smit wrote a note to himself, with a question mark, to celebrate the occasion: "This may be the first evidence of impact (?tsunami) triggered sediment."

Two years later, as sedimentologist Jody Bourgeois loaded up the bags of rocks she collected at the Brazos River site, she had the same thought: Could these rocks be evidence of a tsunami? No one had yet done a study of the sediments, and she was excited to get started. Back in her lab at the University of Washington, Bourgeois gave her samples to a lab assistant, who sliced them up and ground them into thin, nearly transparent pieces. She placed the slices under a microscope to see the size of the mineral grains contained in the samples. If the grains came from the sea bottom offshore, they would be tiny and fine—the makings of mud. If the grains were transported by waves or strong currents, they would contain sand and larger particles. If the sand and mud had been churned up by gentle waves or rhythmic tides, the grains would be organized in a rippled pattern. And if the bottom of the ocean had been disturbed by giant waves, geologic layers could have parts torn out and eroded, and grains as large as boulders might even be moved.

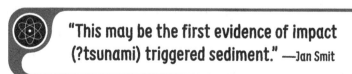

"This may be the first evidence of impact (?tsunami) triggered sediment." —Jan Smit

When Bourgeois placed the slices under a microscope, to her surprise, she found that they were thick with foraminifera, tiny organisms that normally live in the deep open ocean. What were they doing along the Brazos River, an area that during the Cretaceous had connected the Caribbean Sea with the Western Interior Seaway? It looked as if a great wave had pushed deep ocean water through the shallow sea, combining and concentrating foraminifera in an unusually dense mix. Perhaps this, and the mudstone boulder she had seen embedded in the sandstone, was evidence of a tsunami.

Bourgeois researched other Gulf and Caribbean K-T boundary sites and discovered a paper by Haitian American geologist Florentin Maurrasse, one of her mentors at Columbia University's Lamont-Doherty lab. Maurrasse had identified a boundary site in the deep waters near Beloc, Haiti. When he sent a sample of the clay there to the team at the Berkeley lab, Beloc became one of the earliest sites outside of Gubbio, Italy, confirming the worldwide iridium deposits.

Florentin Maurrasse, PhD.

The more she read about Maurrasse's discovery, however, the more Bourgeois realized the similarities between the sediment at the Beloc site and the rocks in the Brazos River. Below the iridium-rich mudstone in both sites lay a bed of sandstone, evidence that turbidity currents—strong, submarine avalanches—had torn up the bottom of the ocean. Turbidity currents leave behind a telltale trace: sediment that has resettled into layers, with coarse sediments on the bottom grading into finer sediments on top. At the time of Maurrasse's discovery, it was not apparent what an important clue was revealed by these layers. Now, alongside Bourgeois's samples from the Brazos River, they seemed like rings around a bull's-eye.

In her lab, Bourgeois leaned over a map with a mathematical compass. With a colleague, sedimentologist Patricia Wiberg, PhD, she had already calculated how big a wave would need to be to move a large boulder. Now they calculated how far away an impact could have been to send a tsunami as far as the Brazos River site. The answer: about 2,000 kilometers. She made a circle with a radius of 2,000 kilometers on the map around the Brazos site. Only a tsunami that started closer than the edge of the circle could have had the energy to reach the Texas site. In 1988, Bourgeois and Wiberg published a paper about their findings and presented their calculations at a conference. The search to find a crater within their circle was on.

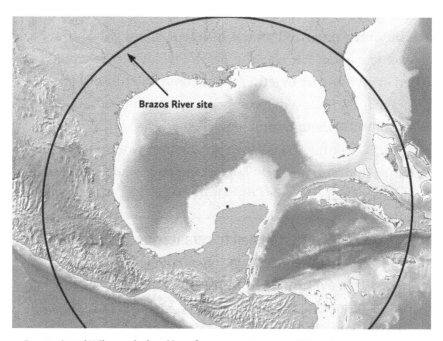

Bourgeois and Wiberg calculated how far away an impact could have been to create waves that reached the Brazos site.

SAD NEWS

Back in Berkeley, Luis contacted Walter with some sad news. Luis had been exposed to a good deal of dangerous radiation in his long career as a physicist. Now he had a brain tumor and esophageal cancer. Over the next year, he sought a cure for his cancer, undergoing multiple treatments and surgeries, but he ultimately succumbed to the illness on September 1, 1988.

After Luis passed away, scientists, colleagues, students, family, and friends gathered to celebrate his life. They published a book called *Discovering Alvarez*, a collection of their memories of Luis and his impact on their lives, paired with the actual research he produced with them. Richard Muller, whose plans to use beryllium-10 to date the Cretaceous boundary rocks fell through, spoke of Luis's sense of adventure in science and his ability to see the interconnectedness of different areas: "Alvarez seemed to care less about the way the picture in the puzzle would look, when everything fit together, than about the fun of looking for pieces that fit. He loved nothing more than doing something that everybody else thought impossible."

After Luis's funeral, Walter got back to work. He still had a crater to find, and he knew Luis would have wanted him to continue the search. That was the Alvarez way.

> "Alvarez seemed to care less about the way the picture in the puzzle would look, when everything fit together, than about the fun of looking for pieces that fit. He loved nothing more than doing something that everybody else thought impossible." —Richard Muller

FIND THE GAP

As Walter kept an eye on research and developments with the crater, he sought out more scientists to lend their expertise and energy to the search. University of Arizona graduate student Alan Hildebrand decided to study the K-T boundary in the Brazos River area as his PhD thesis project. As he combed through the evidence for a tsunami, trying to figure out where a wave might have begun, Hildebrand narrowed his focus from the whole Caribbean to what is now the Gulf of Mexico. The Gulf contained deeper water than the Western Interior Seaway, but it was still close enough to both the Brazos River site and Beloc, Haiti, for a tsunami to churn up sand and move boulders. Using calculations made by Bourgeois and her colleagues, he reiterated her idea that tsunamis from farther away, or from a different direction, would not have had the energy to lift giant rocks. He agreed with Bourgeois—the impact must have happened in the Gulf.

Hildebrand presented these ideas at geological meetings in 1990. After hearing him speak, Walter Alvarez was struck with a new idea, a new way of looking at the evidence. What if they didn't look for what was IN the geological record, but instead looked for what was NOT in the geological record? What if a tsunami had ripped out a piece of rock and created an unconformity—a gap in the record—where neat, orderly strata of rock showed a missing piece? Perhaps that missing piece could point to the tsunami's origin, and the impact site.

Walter began to look for gaps in the geologic record in the **drill cores** of ocean sediment prepared by the Ocean Drilling Project from the Lamont-Doherty lab. Drill cores are long tubes of rock and sediment taken from drilling down below the earth's surface or the seafloor, and they show what is in the rock underground. Perusing hundreds of records from around the world, he discovered that there was only one place that a piece of the strata's sedimentation record was missing—in the Caribbean, at a site called "Drilling Leg 77." Could this be evidence of a crater?

Five libraries of drill cores—samples of rock and sediment pulled up from the bottom of the ocean—are scattered throughout the world, ready for scientists to use to study sediment from the ocean floor and throughout the globe.

MYSTERY IN MEXICO

Hildebrand, meanwhile, took the crater research in another direction: **the gravity fields** of the Gulf of Mexico.

Looking at gravity maps of the Gulf of Mexico, Hildebrand noted a **gravity anomaly** close to -40 in the waters off the Yucatán Peninsula of Mexico. These numbers indicated that a deep crater was hidden in the porous limestone crust somewhere under the Gulf.

Geologists in Mexico working for the government-controlled oil company PEMEX had discovered this gravity anomaly in the

Gravity Anomalies

Everyone experiences the force of gravity. We feel it when we jump up and fall back down. Gravity is a measurable universal force that pulls objects toward each other based on their mass. It also pulls objects on Earth toward the center of the planet. But that force is not equal everywhere. While the difference is so small that it's impossible to feel, the denser the ground is beneath the surface, the bigger the gravity *anomaly*, or difference, that scientists can measure. Sea level generally has a gravity of zero, while the tallest mountains in the world, the Himalayas, have a gravity anomaly of fifty. In places where the ground drops below sea level, or the ground is less dense and more porous, the gravity anomaly will be a negative number, less than zero. Low density can sometimes indicate a rock has tiny holes all through it, with oil hiding in the rock. For this reason, low gravity anomalies interest oil companies. A gravity anomaly of -40 in the Yucatán indicated that there was a crater or extremely porous rock off the coast.

1950s. PEMEX's surveys revealed varying levels of gravity in concentric circles around a central point in the Gulf near Chicxulub (pronounced "CHICK-shoe-loob"), a small town on the northern tip of the Yucatán. Additionally, PEMEX aerial surveys using a **magnetometer**, an instrument that reveals iron deposits, showed an unusual amount of iron for an area composed entirely of limestone. When PEMEX scientists drilled down to check for oil, however, they discovered that the rocks were not oil-bearing but made of andesite, a common mineral in volcanoes.

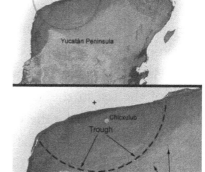

In the 1970s, Glen Penfield, a geologist working for PEMEX, flew over Chicxulub with a magnetometer and found concentric deposits of iron, clear evidence of an asteroid strike. He concluded that Chicxulub was not a volcano, but a meteor crater larger than anything anyone had ever seen.

Some PEMEX geologists concluded that the gravity rings must have been caused by a buried volcano. Glen Penfield, however, believed it must be an impact crater.

Since Penfield worked for PEMEX, his geological work was owned by the company. He gave only one short lecture about his discovery, to a small audience, with geologist Antonio Camargo Zanoguera, and published an abstract, a brief summary of his talk.

Walter Alvarez, Glen Penfield, and Antonio Camargo Zanoguera.

When Alan Hildebrand learned about the gravity and magnetic anomalies at Chicxulub, he contacted Penfield and asked to see the drill core samples he had pulled up from the area. Unfortunately, Alan was told, all the samples from Chicxulub had been destroyed in a fire.

Walter sprang into action. With no access to drill cores, they would have to find another way to determine whether the Chicxulub crater had been created at the time of the K-T extinction. He began looking for sediment that had once been on the floor of the sea in the Yucatán but had now been lifted up to become land, just like the Gubbio limestone he studied in Italy had been lifted from the seafloor to the mountains. He contacted Jose Longoria and Martha Gamper, experts in the micropaleontology of northern Mexico. With their help, he found an area in the mountains and deserts of northeastern Mexico, in the state of Zacatecas, that had once been covered by the Cretaceous seas.

With Nicola Swinburne, a postdoctoral researcher from England, Jan Smit, Sandro Montanari, and Walter's wife, Milly, Walter flew to Mexico in January 1991. As they drove along desolate mountain and desert roads suggested by their Mexican colleagues, Walter kept his eyes open for evidence that would reveal the K-T boundary.

As the days passed, the group drove farther and farther south. With a rickety car motor and return plane tickets a few days later, Walter's team was running out of luck. On their last afternoon, as the sun began to sink into the desert horizon, they bounced down the thick sand of a dry riverbed called Arroyo el Mimbral, looking for outcrops of gray mudstone layered in the sand on the banks of the riverbed. At each appearance of another gray layer, Jan Smit would leap out of the car, check for foraminifera with his hand lens, and call out the geologic time—"Cretaceous!" and "Still Cretaceous!"—moving closer and closer to the K-T boundary.

As the walls of the riverbed cut deeper into the arroyo, and the cloudless sky darkened to a deep gold, an outcrop rose up before them, unlike anything they had seen before. Everyone scrambled out, racing to the rock with hammers and hand lenses. As they examined the layers, they shouted across the riverbed about their discoveries. Everywhere they looked, they spotted spherules and splinters of fossilized wood mixed with deep water sediments full of foraminifera. It was a treasure trove of evidence. An ancient tsunami had clearly ripped through the region, 1,600 kilometers from the impact crater, at the K-T boundary.

At the end of the Cretaceous, this site—now a hot, dry desert—had lain in a deep, calm area of the ocean. The bottom of the ocean was made of clay mixed with the calcium carbonate shells of microfossils such as foraminifera, which would eventually harden into limestone. Now, in the desert, the geologists could see a layer of **marl**, a mixture of clay and limestone that remained from the Cretaceous ocean floor just before the tsunami hit. They identified three distinct layers of sand above the marl. The first and lowest layer contained chunks of ripped-up deep sea bottom combined with spherules and limestone ejecta—chunks of the Yucatán Peninsula that had been catapulted red hot and skyward by the impact and rained down on the site like a blizzard of shooting stars. The second layer looked as if everything on land and sea had been thrown in a blender. Rocks and minerals, splinters of wood from torn-up forests, sand, mud, and shells from coastlines were all smashed and jumbled together, as if tumbled about by the massive tsunami. The third layer showed ripples of clay, sand, and the tiny balls of melted minerals called tektites—evidence of waves sloshing back and forth for days across what would become the Gulf of Mexico. Finally, on top of these deposits, the scientists found the telltale layer of iridium from the impact of the meteor. The iridium must have settled into place over months, drifting down from the atmosphere as the water grew calm again.

As the sun set over the Mexican desert, the scientists rejoiced as they gathered rock samples and set up camp for the night. A few days later, Walter, Milly, and Nicola Swinburne returned to Berkeley with bags of rocks. Sandro Montanari and Jan Smit stayed in the area to collect evidence, photographing and drawing careful pictures of the site. They wanted to capture every detail to provide proof that a massive tsunami had struck the region exactly at the K-T boundary—definitive evidence that Chicxulub, 1,600 kilometers away, was the site of an earth-shattering impact at the time of the Cretaceous extinction.

FINDING A SMOKING GUN

Throughout 1991 and 1992, as Walter and his colleagues analyzed and wrote up the results of their Arroyo el Mimbral study, scientists from all over the world continued to make new discoveries, adding to the evidence that Chicxulub was the crater site. One particular area of interest was spherules from Haiti and the Arroyo el Mimbral site. When minerals are superheated and then cooled instantly in an impact, they turn into a substance called impact-melt glass. The glass reveals the original chemical composition of the minerals and can show whether their source was continental or oceanic crust. Very often, however, these tiny drops of glass turn into clay over millions of years, and the information they contain is lost.

Four groups from four different countries found the same kind of glass spherules as had been discovered in Haiti and Arroyo el Mimbral. Some of them revealed continental crust streaked with minerals from the limestone deposits, such as those along the Yucatán coast. The colors of the glass—black with yellow streaks—showed that the minerals had heated and cooled so quickly that they had not had time to mix together. Spherules from a volcano would have stayed liquid long enough to recombine, so this provided more evidence that an impact, not a volcano, had caused the spherules to form.

European paleontologists visiting El Mimbral—from left to right: Philippe Claeys, Robert Rocchia, Jan Smit, and Eric Robin.

Walter wished they had been able to drill down to get core samples directly from the Chicxulub site, but this was not possible at the time. Then he received good news from Glen Penfield and researchers at the Mexican Petroleum Institute, the research wing of PEMEX. They had found some of the drill core samples from Chicxulub from the 1950s, the ones that were thought to have been lost in a fire. Glen Penfield sent them to Walter and his colleagues, who examined the cores and found impact-melt rocks containing a base of continental-crust minerals mixed with oceanic sediment. This solved an important mystery: whether the meteor had crashed in the ocean or on land. The cores indicated that it landed on an intersection of ocean and continent, creating a confusing mix of minerals from both types of crust.

Over the next two years, scientists from all over the world reexamined the K-T boundary evidence in different ways, challenging each other's findings and confirming or refuting their hypotheses. One study revealed that tektites from the Haiti and Arroyo el Mimbral sites were found at precisely the same layer of strata as the boundary of foraminifera extinctions. This meant that the Cretaceous extinction happened at the same time that the impact created tektites. In another study, scientists from the US Geological Survey used radiometric dating techniques to measure the age of the Chicxulub melt rocks and found their age matched the K-T boundary. And finally, chemical studies of the glass spherules from Haiti and Arroyo el Mimbral revealed that they had the same unusual chemical composition as the core samples from Chicxulub, which meant they were formed from rock melted during the impact.

Newspaper and magazine reporters all over the world flocked to Walter's doorstep to hear more about the discovery. The *New York Times* called the crater and the glass spherules a "smoking gun" for the extinction. The story had all the elements of great drama—dinosaurs, disaster, and death—that made it worthy of banner headlines.

In 1992 and 1993, Walter and over a dozen other scientists published scientific peer-reviewed papers presenting the Chicxulub crater site as evidence of an impact, with arguments that the explosion and aftermath caused the great Cretaceous extinction. It seemed, at last, that the mystery of the extinction of the dinosaurs had been solved.

But in science, every answer is a path to new questions.

DOOMSDAY: The Last Day of the Cretaceous

The last day of the Cretaceous Period, Planet Earth, 66 million years ago.

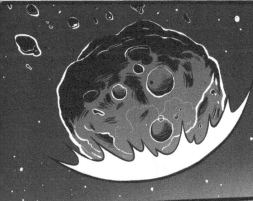

Scientists believe that an asteroid around 10 kilometers (6 miles) in diameter slammed into Earth's atmosphere above what is now the Yucatán Peninsula.

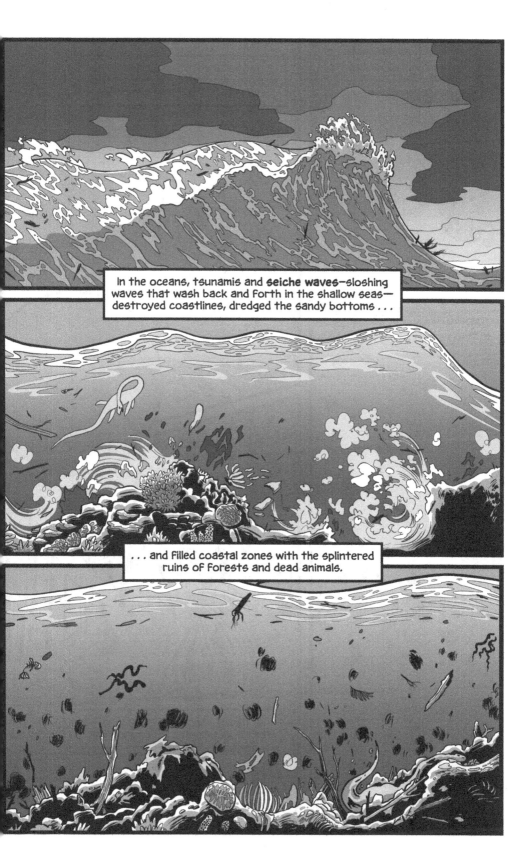

CHAPTER SIX
A NEVER-ENDING STORY

Pincelli Hull, PhD, professor of Earth and Planetary Sciences at Yale University and Associate Curator of Invertebrate Paleontology at the Yale Peabody Museum, has spent countless hours of research aboard the JOIDES *Resolution*. Affectionately called the J.R., the JOIDES *Resolution* is a science research vessel capable of drilling down through the ocean floor. Hull and a team of colleagues use the J.R. and other ocean drilling vessels to discover information about the ancient past written in the sediment at the bottom of the ocean. This dull, gray, sandy soil holds the key to understanding the climate, air quality, and organisms that lived in Earth's ancient past.

Scientists on the JOIDES *Resolution* drill into the ocean's core to find geological evidence.

Despite the abundance of evidence that an asteroid impact happened at Chicxulub at the time of the Cretaceous extinction, scientific exploration is never complete. Science is a process, not a product or even a project that has a beginning, middle, and end. Science is about making observations, asking questions, creating hypotheses, and finding ways to test and experiment to find the truth. And when one question is answered, a new one is always waiting in the wings.

While most scientists agree that a giant asteroid hit the earth at the end of the Cretaceous Period, new questions continue to surface: Did the impact cause the extinction, or did it just happen at the same time? If it was the cause, exactly how did the impact lead to the death of over 75 percent of all species? Did anything else contribute to the disaster? And finally, could it happen again? Hull and other scientists on the J.R. are constantly designing new experiments to narrow down the details of what is now called the K-Pg extinction. (The name of the Tertiary [T] Period was changed to the Paleogene [Pg] in 2008, hence the new name.)

Hull has said she thinks of the JOIDES *Resolution* as a time machine that allows her to see Earth's deep past. "If you take the J.R., the boat itself, and you add to it all the scientists and technicians that are on board, you can think of the J.R. as a true time machine," she says in a video. "As you drill into the ocean bottom, you are drilling back in time."

Core drill bit used on the JOIDES *Resolution*.

To date, the J.R. and other vessels like it have drilled into hundreds of sites all over the world, pulling up drill cores—long tubes of sediment—and cataloging their locations. These tubes are then opened up, sawed in half lengthwise, sampled, and examined using chemistry, magnetic instruments, and microscopes to identify the time signature of each layer of strata. Each tube of sediment offers a picture of the past.

Hull says, "If you come into the core lab, and we open one of these cores... around one of these really interesting time periods where the world is changing... you see the world change before your very eyes."

In one sample, the tube of mud changes abruptly from white to brown. The white sediment shows a healthy ocean full of foraminifera and other microorganisms. The dark brown mud is lifeless, the result of ocean acidification, when the shells of foraminifera dissolved and mass extinctions plagued the planet. The change is evidence of a time when Earth's temperature warmed dramatically, and it may provide hints of our own future under climate change.

> "You see the world change before your very eyes." —Pincelli Hull

Ocean Acidification

Earth's atmosphere maintains a delicate balance between oxygen and carbon dioxide in order to support life. Animals breathe in oxygen and breathe out carbon dioxide, while plants take in carbon dioxide and release oxygen. A healthy atmosphere supports life in the ocean too. When the atmosphere is in balance, the ocean takes in about 30 percent of Earth's carbon dioxide, releasing oxygen.

Sometimes, natural disasters and human activities pollute Earth's atmosphere with more carbon dioxide than the oceans can absorb. The atmosphere becomes a blanket around the earth, causing the planet to heat up. The extra carbon dioxide in the atmosphere also combines with water to become carbonic acid, which dissolves the shells of foraminifera and other marine animals. Scientists can tell how much acid was in the ancient oceans by measuring the pH, or potential hydrogen, of ocean sediment. Normal seawater has a pH of 8.1. A low pH in ancient oceans indicates that there was a lot of carbon dioxide in the air, a very warm atmosphere, and acid in the ocean. Learning about the pH levels of the ocean during past warm periods can give us clues not only about what happened in the past, but also what to expect as climate change warms the planet today.

In her lab at Yale University, Hull inserts a slide into a microscope. Her fingers hover over a row of round black window frames, each the size of a dime. The little windows are full of tiny white specks that look, to the untrained eye, like grains of sand. Hull dips a fine-tipped paintbrush in a bottle of water, lowers her head over the microscope, and peers through the lens at the delicate shells of her favorite organisms in the world—foraminifera. With the microscope magnifying the shells hundreds of times, Hull examines the curving structures of the ancient animals and picks them up with her wet brush.

To her, these aren't just tiny fossils; they are records of Earth's history, written with the chemicals in their shells. Together, they are like tiny libraries, "telling stories about past mass-extinction events, past abrupt climate change events, and essentially how the earth has evolved over really long periods of time." The foraminifera are clues in her ongoing quest to gather data about the chemical signature left behind from the Cretaceous extinction. Decades after Walter first presented his theory regarding an asteroid impact, Hull is hot on the trail of new evidence about how the Chicxulub impact caused the dinosaurs to die out. She is one of many researchers continuing to probe the earth to confirm—or perhaps disprove—the Alvarez hypothesis.

THE HEART OF CONTROVERSY

Hypotheses are made to be questioned, challenged, confirmed, or disproven by data. That is the scientific process. The more data scientists have, the more ideas can be verified or disproven, and the more questions crop up. Scientists from a variety of disciplines continue to find evidence about what caused the Cretaceous extinction and why it is important to us today.

While nearly everyone agrees that the Chicxulub impact happened at the K-Pg boundary, a few geologists still wonder whether it was the *only* cause of the Cretaceous extinction, or whether it was the last in a series of cascading disasters that ended much of life on Earth. Gerta Keller, PhD, professor of Geological and Earth Sciences at Princeton University, still believes that volcanism—volcanoes—played a role.

In a 1988 paper, Keller published her analysis of evidence she had collected over a three-year period at a K-T boundary site in El Kef, Tunisia, deep in the North African desert. Her work there suggested that the Deccan Trap volcanoes in India sent so much lava, carbon dioxide, and sulfur into the atmosphere that it caused the climate to change 300,000 years before the Cretaceous extinction, sending many foraminifera species into a tailspin of population decline. Speaking before a large audience at the second Snowbird conference, Keller proposed that a single impact could not have caused such a drastic reduction in foraminifera. Perhaps there was more to the extinction than a single catastrophic event, she said. Perhaps the volcanoes were involved.

Jan Smit raised a finger to disagree with Keller's analysis of what the evidence suggested. He had been examining the same data set—taking samples of the very same foraminifera—and did not find a single extinction in the 300,000 years Keller's evidence suggested it had happened. Other scientists in the crowd had less polite ways of voicing their opinions. The exchanges between these scientists and Keller unleashed another chapter in the "nastiest feud in science," a battle between uniformitarian and catastrophist views of the extinction. Keller and other twentieth-century uniformitarians interpreted their data to mean that volcanoes had killed off dinosaurs over a long period of time, while the catastrophists saw evidence that the Chicxulub asteroid had triggered a rapid extinction. While disagreement is always part of science, the dispute steeled Keller's determination to keep looking for evidence of the volcanoes' involvement in the extinction.

Keller says her interest, like those of other scientists, has always been to find out why mass extinctions happened. "Why do certain species go extinct and not others? I work my way back into history. I like to call myself a science detective of very old cases. It is cold case detective work, finding out what happened 66 million years ago."

Keller doesn't question that a massive asteroid hit the earth. There is irrefutable evidence of an impact. But she thinks there is more to the story. "The impact happened," Keller says. "But so much evidence cannot be explained with that."

Like Hull and Smit, Keller identifies species of foraminifera to see what was living at the time their shells fell to the bottom of the ocean. "We collect samples in the field, bring them back to the lab, and wash away sediments with a fine sieve until we have nothing but very clean foraminifera shells," she explains. Some species lived only when ocean conditions were perfect—the right temperature and pH.

Some species of foraminifera could only live when oceans were cool, while others survived warmer conditions.

Other species were survivors. They thrived and expanded under bad conditions. The extinction rates of foraminifera species, and the increased presence of these survivor species, provide clues about whether the environment was too warm, and full of smoke and acid rain, or healthy.

Keller also measures the percentages of different isotopes of oxygen and carbon within the rock and foraminifera shells to calculate the ocean temperature during a period in geologic time. Different isotopes of carbon can indicate whether the water was cold and full of nutrients, or warm and less able to support life. Ultimately, Keller's analysis of data concluded that pulses of volcanic flows led to atmospheric changes, which resulted in foraminifera—as well as dinosaur—extinctions.

UNLOCKING THE DETAILS

Pincelli Hull has made it the focus of her research to determine whether volcanism or the impact—or both—might have caused the extinction. She has looked for ways to compare the two models: one, that volcanoes caused the dinosaurs' extinction; the other, that the asteroid impact caused the extinction. Rather than looking at the timing of volcanic lava flow, however, Hull's research measures different chemicals in foraminifera shells and sediment to reveal the substances released by volcanoes in a process called "outgassing."

"It turns out what a volcano does to kill off life is release gas," Hull explains. Meteor impacts and volcanism release different gases from the earth's core, which cause acid rain. Acid rain makes the ocean more acidic, dissolving foraminifera shells and adding

Diversity in US Geosciences

Geologists and paleontologists who study the Cretaceous extinction spend a lot of time studying biological diversity—how many different kinds of animals and plants lived before and after the impact. Biological diversity is a sign of a healthy ecosystem. The greater the number of different kinds of plants, animals, fungus, bacteria, and other living things in an ecosystem, the more likely it is that the ecosystem can avoid collapse during times of stress.

Human diversity—including people of different races and genders, as well as different life experiences, perspectives, skills, beliefs, ideas, and ways of life—can also help a society survive and thrive in tough times. The more creativity we can bring to the table, the more likely it is we can think our way out of our toughest problems.

That's why some scientists are alarmed to see a lack of diversity in the geosciences—the study of the earth, atmosphere, and oceans. For centuries, geology has been dominated by scientists who are predominantly white and male. Just as they did in Mary Anning's day, women and people of color continue to struggle for inclusion in the geosciences in the US. Now geologists at the highest levels are talking about this issue and what they can do to change it.

In 2018, researchers Rachel Eleanor Bernard from the University of Texas and Emily H. G. Cooperdock from the University of Southern California investigated how many geoscience PhDs have been awarded in the US to women and non-white minorities in the past forty years. In a study involving both citizens and permanent residents of the US, they found that while PhDs awarded to women and non-white minorities have increased, their numbers still fall short of their population percentages in American society.

"Increased diversity has clear benefits for scientific advancement," Bernard and Cooperdock wrote. "Different perspectives and life experiences spark unique questions and approaches to problem-solving. Collaborations involving a diverse group of people are more creative at tackling problems and lead to higher levels of scientific innovation."

What can be done to increase diversity in geoscience? Kuheli Dutt, PhD, a researcher at the Lamont-Doherty Earth Observatory, thinks the key is for individuals working at research institutions and universities to recognize their biases—the tendency to give preference to a particular group. These biases can make it challenging for women and people of color to advance in the sciences. Who, for example, gets invited on research trips or to conferences, or is mentored by senior scientists? Whose project will be awarded a grant? Making room for diverse voices will bring more ideas—and discoveries—in our quest to understand the earth.

carbon to the ocean. The chemicals in different layers of a drill core, therefore, can show when forams died out and what killed them—volcanoes or an impact.

In one study, Hull and colleagues measured the pH of sediments and shells to find out if the ocean had more acid at the K-Pg boundary, or if pH levels and chemicals were timed to volcanic emissions. Hull's group concluded that the amount of chemicals released over time by the Deccan Trap volcanoes in India would not have been enough to cause a mass extinction, and that a worldwide extinction of foraminifera happened almost exactly at the K-Pg boundary. Looking at all the data, Hull says, "allowed us to say, no, it really was the impact" that caused the extinction of the dinosaurs, and not volcanoes.

DRILLING THE IMPACT SITE

While Hull researches the chemical signature of the extinction in the shells of foraminifera, other scientists are looking at other kinds of evidence to trace what happened when the asteroid struck and how the earth recovered.

Christopher Lowery, PhD, professor of Paleontology at the University of Texas, has drilled into the crater itself—something that was not technologically possible for Walter to do when the crater was first discovered. Working with a team of scientists at the Chicxulub site, Lowery used the *L/B Myrtle*, a "lift" boat with feet that drop down to the ocean floor for stabilization, to bore through shallow water into the sediment at Chicxulub.

Christopher Lowery, PhD.

The drill cores Lowery's group recovered from the crater show layers of rock churned and disrupted first by the impact, then by a tsunami and sloshing seiche waves. When the sediment finally settled down to the ocean floor, it piled up in layers, from large rocks to minute grains of sand and silt on top in a "transition zone," capped by white deep ocean limestone. These sediments tell the story of doomsday.

"What we think happened is visible in the transition zone, the area of sediment that marks the end of the Cretaceous and the beginning of the Paleogene," Lowery said. "Within an hour of impact, water rushed back in, tsunamis

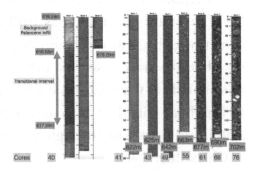

Drill cores show the layers of sediment that filled in the crater, over hundreds of thousands of years.

sloshing back and forth. . . . Then, slowly, as the energy subsided, all the sediment settled down."

Lowery used radiometric dating and the chemical signature of the crater sediment to tell approximately how much time passed during the crater's recovery. Lowery measured helium-3, an isotope of helium (He) that falls from cosmic dust at a known rate. He discovered that the low He-3 rate indicated that fewer than 1,000 years passed before some recovery had begun at the site—an instant in geologic time.

Most surprising, the crater drilling project revealed evidence that some forms of life, like bacteria, burrowing animals, fish, and shellfish, moved back into the crater site relatively quickly, tunneling into the sediment in some parts of the crater while other areas still seeped scalding water from boiling hydrothermal vents.

Lowery believes that the crater site recovered faster than some other parts of the planet because it lay in an area that the ocean washed through freely, constantly bringing in new oxygen and new life. In areas farther away from the crater site, however, his chemical evidence shows that it was 1.8 million years after the impact before Earth's oceans were restored to health and productivity. By then, the Age of Mammals had begun.

Other scientists are also examining sediment from the drill cores to find evidence of the asteroid itself. Sean Gulick, a research professor at the University of Texas Jackson School of Geosciences, and Joanna Morgan, a professor at Imperial College in London, have pinpointed the layer of iridium within the drill core to the time of impact—a "fingerprint" of the asteroid that hit the earth. "It puts to bed any doubts that the iridium anomaly is not related to the Chicxulub crater," according to Gulick.

Other scientists are working to discover where exactly the Chicxulub asteroid may have come from. "Two critical [questions] still unanswered are: 'What was the source of the impactor?' and 'How often did such impact events occur on Earth in the past?'" said William Bottke, PhD, part of a team of three researchers from the Southwest Research Institute Department of Space Studies in Colorado. Bottke and colleagues David Nesvorny and Simone Marchi probed Chicxulub drill cores to determine the chemical composition of the object that hit Earth. They discovered it was most likely part of a class of meteors called carbonaceous chondrites—dark minerals created early in the life of the solar system.

To find where the asteroid came from, the scientists "decided to look where the siblings of the Chicxulub impactor might be hiding," according to Nesvorny. Using powerful NASA supercomputers, the scientists modeled the orbits of asteroids to identify the probable path of the Chicxulub impactor. They discovered it most likely came from the outer part of our solar system's asteroid belt, between Mars and Jupiter. This area contains an estimated 1.1 to 1.9 million asteroids one kilometer in diameter or larger, plus countless smaller objects, many with the same chemical composition as the Chicxulub meteor. Nesvorny estimates that asteroids large enough to have caused the Chicxulub impact could potentially reach Earth as often as every 250 million years.

DINOSAUR DISCOVERIES

While many scientists study microscopic foraminifera to understand what happened at the K-Pg boundary, some are also working at the other end of the size scale, digging up dinosaurs and other large fossils.

Robert DePalma, a paleontology PhD student at the University of Manchester, who has worked at the University of Kansas's Biodiversity Institute and Natural History Museum, has spent a lot of time in the last few years looking at bones.

Lying on his side in a dusty patch of earth in the Hell Creek formation of North Dakota, he picks away at wafer-thin layers of soil with the slender tip of a knife blade, then lifts the mudstone away, one layer at a time. When the soil reveals nothing, he brushes it off with a paintbrush. But whenever he finds something—a fossil leaf, a seed, fish bones, or even the impression of rough skin or feathers—DePalma stops, squirts it with a glue called PaleoBOND, which is used to preserve fossils, and waits for it to soak in. His work, like the work of every paleontologist, is slow and painstaking, because one false move can destroy precious evidence—evidence that can help tell us what exactly happened around the world on doomsday.

DePalma is the principal investigator at Tanis, a site some scientists believe may go down in history as the richest K-Pg boundary fossil collection ever discovered.

Jan Smit and Robert DePalma at Tanis. Some scientists say that Tanis may go down in history as the richest K-Pg fossil site ever discovered.

DePalma began searching for fossils in the Hell Creek formation in 2004, when he was a twenty-two-year-old undergraduate paleontology student. In 2012, he received a tip from an amateur fossil collector who had discovered a site full of delicate fish fossils and mud so soft it could be dug with a shovel rather than a pick. The collector had abandoned the area as not worth his while and told DePalma to dig in. DePalma contacted the landowner for permission, and the next summer he set to work.

From the first, DePalma could see there was something strange and special about the place. Immediately, he discovered what looked like a hailstorm of tiny round specks, or microtektites—glass droplets formed when an impact or explosion melts rock and it falls to Earth as glass rain. As he dug deeper, he found a treasure trove that included whole fish positioned upright in the mud with microtektites stuck in their gills, dinosaur skin and bones, feathers, "logjams" of trees, and amber—fossilized tree sap—with microtektites stuck inside. While the small site is only three feet deep and about the length of a small public swimming pool, it contained a jumble of hundreds of fossils. Strangest of all was the mixture of specimens, with plants and animals that could not possibly live in the same place embedded side by side: freshwater fish and giant marine reptiles, trees, insects, and even the burrows and bones of tiny Cretaceous mammals.

Fossil of paddlefish at Tanis.

While DePalma had first wondered if it was worth excavating the site, he now wondered if he had found the holy grail of dinosaur discoveries: a site that showed what happened on doomsday, in the hours following the Chicxulub impact. Eager to share what he had found and get help verifying his discoveries, he asked the most respected scientists in the field—Walter Alvarez, Jan Smit, Florentin Maurrasse, and others—to visit the site or assist in his research.

Walter was amazed by what he saw. "When we first proposed the impact hypothesis to explain the great extinction, it was based on just finding an anomalous concentration of iridium—the fingerprint of an asteroid or comet," Alvarez told a reporter. "Since then the evidence has gradually built up. But it never crossed my mind that we would find a deathbed like this."

Walter Alvarez and Robert DePalma at Tanis.

Jan Smit agreed. On his first trip to Tanis, he witnessed an astonishing array of fossils. "There are eggs of pterosaurs—flying dinosaurs—with the embryos inside the eggshells. In order to have embryos of a dinosaur embedded in the tsunami, you know [the pterosaurs] were alive at that moment. Nobody knew that. They were supposed to be extinct a million years before that." Smit described other fossils DePalma found at the site: "He found two sets of dinosaur footprints . . . on the banks of a river. The footprints are cast by the tsunami, so you realize those dinosaurs were walking maybe an hour before the tsunami hit." Smit also saw dinosaur teeth, skin, and a long dinosaur feather. The site was mind-boggling, even for someone who had spent a lifetime studying fossils. "I assure you," Smit said later in an interview, "it's all true." On seeing the fish with microtektites stuck in their gills, Smit commented, "This was the first time we had seen a direct victim of the impact."

To find out whether the site truly coincided with the Chicxulub impact, following Jan Smit's advice, DePalma sent a sample of microtektites from Tanis to a lab, along with samples from a known K-Pg boundary site for comparison. He discovered that the chemical composition of the tektites was the same, which meant they were created by the same impact.

"This was the first time we had seen a direct victim of the impact." —Jan Smit

Walter Alvarez and Dr. Mark Richards, an earth science professor at the University of Washington, consulted with DePalma on the timing of the event. When and how could waves have arrived at the site in a manner that preserved both the fossils and the impact tektites? A tsunami would have taken ten to twelve hours to reach the site from the Yucatán, whereas the tektites would have rained down only forty-five minutes after impact. How could fish embedded in the site by a tsunami have gotten tektites stuck in their gills?

Walter and Richards suggested seismic waves. The asteroid could have caused a giant earthquake, of a magnitude of 10–11 on the Richter scale, that traveled through the earth and arrived in Tanis within ten minutes of the impact.

Such a large earthquake could have caused seiche waves to slosh back and forth in the shallow sea that lay over the site, trapping animals in mud and debris just after the tiny glass spheres

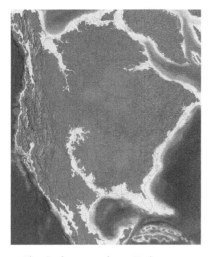

The site known today as Tanis was once located on the shallow Western Interior Seaway.

rained down from the sky. These waves could have mixed freshwater and marine environments, pouring silt and mud into the shallow sea, and suffocating animals caught in the mudslide. This scenario would explain the mystery of why freshwater fish and oceanic fossils—animals that live in vastly different environments—have been found side by side, perfectly preserved.

Such seismic waves have happened before; for example, the 2011 Great East Japan earthquake, of magnitude 9.0, triggered seiche waves in a Norwegian fjord thousands of kilometers away.

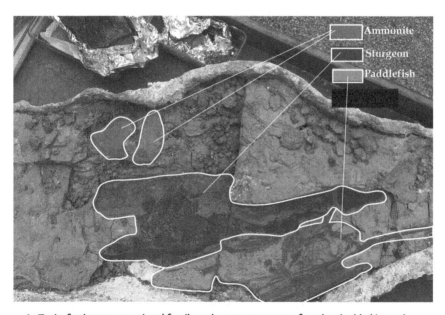

At Tanis, freshwater organisms' fossils such as sturgeon were found embedded in mud next to marine animals such as ammonites.

Walter was delighted with the opportunity to help DePalma develop a scenario for what might have happened on what the *New Yorker* magazine called "the day the dinosaurs died." He and Jan Smit signed on as two of DePalma's coauthors on a scientific article titled "A Seismically Induced Onshore Surge Deposit at the KPg boundary, North Dakota," published in the prestigious *Proceedings of the National Academy of Sciences* journal in 2019.

THE SEASON OF THE IMPACT

Since his discovery, DePalma has invited a number of researchers to work on the site. Melanie A. D. During, PhD candidate from Uppsala University in Sweden, came to Tanis with a question. She wondered if the life stages of fish found in Tanis would offer clues about the time of year the asteroid hit.

From her research, During knew that many fish add layers to their bones seasonally in much the same way that trees add rings each year, growing in warm weather when food is abundant, and stopping growth in cold seasons when food is scarce. She also knew that the chemical signatures of food sources would be evident in the fishes' bones. "Fish generally contain more carbon-12," she explained. "But carbon-13 rises in the summer when they ingest more cyanobacteria, which produces C-13."

During measured the bones, teeth, and spines of six paddlefish and sturgeons from Tanis for biochemical content and evidence of growth. Her evidence revealed that the fish were in the early stages of a spring growth season when they were instantly buried in sediment.

The season the asteroid hit would have been a factor in determining which animals survived and which became extinct in the northern and southern hemispheres. In a paper titled "The Mesozoic Terminated in Boreal Spring," published in *Nature* in 2022, During suggests that in autumn in the southern part of the planet, some animals, such as mammals, some amphibians, and crocodilians, could have still been hibernating in burrows underground, protected from fires that raged following the impact. This may explain why these groups of animals survived the impact. At the same time, in the northern, or boreal spring, on the other hand, species that had to incubate eggs for long periods of time, such as dinosaurs and pterosaurs, could have been exposed to the cold and would not have survived.

In researching the impact, other scientists are looking at the chemical composition of rocks for evidence of climate-changing sulfur. Christopher Junium, Aubrey Zerkle, and a team from the United States, Scotland, and England have determined that fine particles of sulfate soared beyond the atmosphere into the stratosphere, blocking the sun and causing a prolonged "impact winter" that resulted in mass extinctions. "The unique fingerprints we've measured in these impact sediments provide the first direct evidence for the importance of sulfur aerosols in catastrophic climate change and cooling," Zerkle said in an interview with Syracuse University. Junium explained that "sulfur aerosols would have extended the duration of post-impact climate change," leading to the end of major life systems on Earth.

BIG SCIENCE, BIG HISTORY, BIG LIVES

While newly minted PhD scientists continue to research the impact, the scientists involved in uncovering the original clues are still at work, more focused now on helping younger scientists than pursuing their own research plans. Jan Smit consults on a variety of projects, from examining Christopher Lowery's drill cores, to introducing Pincelli Hull, Melanie During, and even the author of this book to his favorite K-Pg boundary sites, to supporting Robert DePalma as he uncovers evidence at Tanis. Jody Bourgeois has been honored as a Fellow of the Geological Society of America for her contributions to the field. While she retired after thirty-six years of teaching at the University of Washington, she continues to do field research in Siberia and consult with the next generation of geologists as they uncover evidence of paleotsunamis. In her spare time, she is passionate about researching women pioneers in the field of geology. She wants the world to know about centuries of women who have made an impact in science.

As for Walter, these days he lets the science—and scientists—speak for themselves. Since 2006, he has devoted much of his time to a project called "Big History," teaching a class that ties the history of the people of Earth to the larger history of the planet and universe. While he has visited and weighed in on the work of Robert DePalma at Tanis, he largely leaves the continuing battle over what caused the dinosaurs' extinction to others. What he wrote in 1997 in his book *T-Rex and the Crater of Doom* still applies today: "Right now, we are in a situation that scientists particularly enjoy—where there is an intriguing mystery, some obviously significant clues, and nobody has an idea what the explanation will be."

Walter was sorry that his father, Luis, didn't live to see the discovery of the crater. But he was glad for the chance to work with him on one of the biggest discoveries in the history of the earth. Together, they followed the path set out by generations of Alvarez scientists—keeping their minds open and curious, taking time to be quiet and see connections, and not being afraid to take risks.

Luis and Walter Alvarez posing near one of the K-Pg boundary outcrops in Italy that first caused Walter to wonder about the meaning of the line of fossil-free mud.

In *T-Rex and the Crater of Doom*, Walter wrote, "[Dad] had delighted in the effort to get past all the tricks and stumbling blocks Nature had placed in the way of finding the site of the impact. Dad would have loved the discovery of the Crater of Doom."

So what really happened at the end of the Cretaceous? Scientists agree that a massive asteroid ten kilometers (six miles) in diameter struck near Chicxulub, Mexico, and caused one of our planet's great extinctions—as noted earlier, it wiped out over 75 percent of species and 99.99 percent of living things on Earth. But questions remain. How did the Deccan Traps volcanoes impact life on Earth before the asteroid hit? How exactly did the extinction unfold? How did Earth recover? And what can we learn from the post-impact climate change that might help us understand our currently changing climate?

Scientists—geologists, paleontologists, physicists, and every other researcher in every field—never stop digging for data, collecting evidence, and retesting results. The scientific practices of asking questions, sharing ideas with colleagues from around the world (who often become friends), developing a hypothesis, and testing to see if it is actually true are the eternal pleasures of doing science. Science presents a never-ending opportunity to learn and grow, and to expand the knowledge of civilization. Scientists don't work for themselves; they work for everyone—*everyone* on the entire planet, now and in the future—to bring the truth to light and make it accessible and available to the whole world.

What we find out about the Cretaceous extinction may help us prevent what scientists say is a disaster sitting right on our doorstep—potentially the sixth great extinction, caused by human-made climate change. Perhaps you, reading this book right now, will be the scientific detective who uncovers another magnificent truth, untangles a science mystery, finds the solution to a perplexing puzzle, and opens the doors of science wide, so everyone can stand in the light.

GLOSSARY

asteroid – a space object or minor planet, too small to be considered a planet, but too large to be a meteor.

astrogeology – the study of the geology of space.

astronomer – a scientist who studies space.

astrophysics – the study of the physical nature and behavior of space objects as large as stars and planets, and as small as cosmic rays.

cartographer – a person who makes maps.

catastrophism – the idea that geological changes on the earth happen quickly, through catastrophic disasters.

catastrophist – a person who believes in catastrophism.

classify – to arrange a group of people or things in categories according to shared qualities or characteristics.

continental crust - the crust of the earth on land, including silicon, quartz, feldspar, and pyroxene.

continental drift – Alfred Wegener's theory that the continents are not stationary.

convection currents – the circular movement of liquid or gas such as magma or air, which heats up, rises, moves out, cools, and falls again. Convection currents serve as the engine for plate tectonics, as well as for the water cycle, weather, and climate.

cosmic dust - dust and small debris from space that includes elements such as iridium that are rare on the surface of planet Earth.

cosmic rays – energetic subatomic particles traveling through space at light speed.

crust – the top layer in the surface of planet Earth.

cyclotron – a machine in which charged subatomic particles are accelerated to help scientists understand their structure and behavior.

deposits/deposition – the placing down and layering of soil or sand by wind or water.

drill core – a cylinder-shaped section of Earth formed by drilling a hollow tube down into the ocean floor or Earth's crust to see what is layered under the surface.

ejecta – substances that are thrown out of a volcano or impact crater.

electroscope – an instrument for detecting electric charge.

erosion – the wearing away of a surface, such as soil or sand, by wind or moving water.

evolution through natural selection – the gradual process by which species change over time in response to environmental conditions.

extinction – for a species to no longer exist.

foraminifera (forams) – various species of microscopic single-celled organisms with shells made of calcium carbonate and other materials often used to identify periods of prehistory.

gamma rays – electromagnetic radiation that comes from the nucleus of radioactive elements emitting particles.

geology – the study of the earth, including rocks and minerals.

gradualism/gradualist – the belief that the earth changes gradually and slowly; a person who believes in gradualism.

gravity anomaly – an unexpected difference in the gravitational energy of a place.

gravity field – the gravitational energy of a place.

half-life – the time it takes for an isotope of an element to lose half of its unstable particles.

hiatus – a period of time missing from the geological record, visible as an unconformity in the rocks.

hypothesis – a proposed explanation for a phenomenon based on evidence.

igneous – rocks made from fire, or melting, such as volcanic rocks or granite.

impact crater – a crater created by the impact of an asteroid or meteor.

ion – an atom or molecule with an electric charge due to an unbalanced number of electrons and protons.

ionizing radiation – radiation that causes atoms to lose electrons.

isotope – a form of an element with a different number of neutrons in its nucleus.

K-T boundary/K-Pg boundary – the boundary between rocks from the end of the Cretaceous Period (K is from the German spelling of "Cretaceous") and the beginning of the Tertiary (T) Period marking the extinction event that killed 75 percent of species on Earth. The name of the Tertiary was changed to Paleogene (Pg) in 2008 to become the K-Pg boundary.

Law of Superposition – the law that states that when sediments are deposited, the oldest one will be on the bottom, and the newest one will be on the top.

lithosphere – the semi-liquid outside layer of the mantle that Earth's crust floats on.
magma – melted rock inside the Earth.
magnetic polarity – the state of Earth in which there is a North Pole and a South Pole with magnetic attraction and repulsion, and the alignment of iron-rich crystals in rocks toward the dominant pole.
magnetic reversals – a change in Earth's magnetic polarity from the North Pole to the South Pole, visible in the changed alignment of iron-rich crystals in rocks.
magnetometer – an instrument that detects the presence of iron in rock.
mantle – the semi-liquid, melted layer of magma surrounding Earth, under the crust, above the solid core.
marl – a mixture of clay and limestone.
metamorphic – rock formed when igneous or sedimentary rock is remelted and reformed under heat and pressure.
meteor – a small object from space that enters earth's atmosphere.
meteorite – a meteor that reaches Earth.
meteorologist – a scientist who studies the atmosphere, weather, and climate.
mid-ocean ridge – mountain ranges running through the middle of the earth's oceans at plate boundaries, where magma emerges from the seafloor.
natural theology – the eighteenth- and nineteenth-century understanding of God through observation of natural occurrences.
oceanic crust – the earth's crust at the bottom of the ocean, often including basalt composed of silicate and magnesium.

paleontology – the study of fossil animals and plants.

peer-review process – the process by which scientists read and review the papers of other scientists before publication to determine whether their experimental process was accurate and reliable.

PhD – a Doctor of Philosophy degree is the highest degree that can be earned in a university setting and can be awarded in most non-medical and non-legal fields, including disciplines of science. Most professors and scientists have earned PhD degrees.

physics – a science that studies the structure and behavior of matter.

physiology – how the bodies of living things function.

plate tectonics – the movement of plates in the earth's crust.

radiation – the emission of subatomic particles in elements seeking stability due to an imbalance in the number of protons and neutrons.

radioactive – emitting radiation, or subatomic particles.

rift – a canyon or valley formed when tectonic plates move apart.

scientific method / scientific practices – the practices used to answer a scientific question or solve a scientific problem.

seafloor spreading – the mechanism for plate tectonics in which magma pushes out of a rift in the seafloor and pushes plates in the earth's crust apart.

sedimentary – rocks made from tiny particles of minerals carried by wind or water, deposited in a place, and pressed together by pressure.

seiche [saysh] **wave** – a standing wave that sloshes back and forth in a partially enclosed body of water; often caused by an earthquake or other geologic disturbance.

shocked quartz – quartz that has been subjected to extremely high pressure due to an impact, resulting in a distinct pattern of cracks.

sonar – an instrument that bounces sound waves off objects such as the ocean floor to make shapes visible.

spherules – small spheres of molten rock formed from minerals in the vapor released after an impact, which cool in the atmosphere and fall to Earth.

Stokes's Law – the law in physics that explains that the speed at which a round shape will move through fluid depends on the size of the sphere and the thickness of the fluid.

strata – layers of rock.

stratigraphic – relating to the order and organization of layers of rock.

subatomic particles – parts of an atom.

subduction zones – the edges of plates in the earth's crust where one plate is pushed underneath another plate, where it is remelted in the mantle.

supernova – the explosion of a star; a high-energy event that emits radiation.

tektites – small spheres of melted volcanic glass thrown skyward by an impact.

unconformity – a break in the geologic record through erosion with evidence that strata are missing.

uniformitarianism – the idea that the slow, regular processes that shape Earth today also shaped it in the past.

uplift – the rise in elevation and formation of mountains caused by two tectonic plates pushing together.

SOURCE NOTES

THE BIG EVENT

This is a story about how curiosity, scientific processes, and cooperation across disciplines can answer universal questions and solve earth-shattering problems that benefit the whole world. To tell this story, I needed to understand and weave together discoveries from different areas of science—paleontology, astronomy, physics, geology, sedimentology, biochemistry, atmospheric chemistry, and oceanography, to name a few—with the very human stories of the scientists themselves, to create a single narrative.

The following resources, organized by category, helped me understand the science, the people, and the stories that came together to answer one big question: What killed the dinosaurs?

THE BIG PICTURE

Walter Alvarez's 1997 book, *T-Rex and the Crater of Doom*, and Luis Alvarez's 1987 book, *Alvarez: Adventures of a Physicist*, helped me create a narrative throughline.

Alvarez, Luis W., *Alvarez: Adventures of a Physicist*, Basic Books, New York, 1987.

Alvarez, Walter, *T. Rex and the Crater of Doom*, Princeton University Press, Princeton, NJ, 1997.

EARLY DINOSAUR DISCOVERIES AND IDEAS OF EXTINCTION

Benton, M.J., "Scientific Methodologies in Collision: The History of the Study of the Extinction of the Dinosaurs," *Evolutionary Biology*, 1990.

Buckland, William, *Geology and Mineralogy Considered with Reference to Natural Theology*, vols. 1–3, Pickering, London, 1836.

Buckland, William, "Notice on the Megalosaurus or Great Fossil Lizard of Stonesfield," in *Transactions of the Geological Society of London*, February 20, 1824, pp. 387–398.

Buckland, William, *Reliquiae Diluvianae or Observations on the Organic Remains Contained in Caves, Fissures and Diluvial Gravel and Other Geological Phenomena Attesting to the Action of a Universal Deluge*, John Murray, London, 1823.

Cadbury, Deborah, *The Dinosaur Hunters,* Fourth Estate, London, 2000.

Conybeare, Rev. W.D., "On the Discovery of an Almost Perfect Skeleton of a Plesiosaurus," *Transactions of the Geological Society of London*, February 20, 1824, pp. 381–388.

Cuvier, Georges, "Mémoires sur les espèces d'éléphants vivants et fossiles," 1800, http://biodiversitylibrary.org/page/16303001.

Cuvier, M. le Baron, "Discourses on the Revolutionary Upheavals on the Surface of the Globe," *Fossil Bones*, Paris, 1825, www.victorianweb.org.

Historical Weather Records: Weather Web, https://premium.weatherweb.net/weather-in-history-1800-to-1849-ad/.

Ketchum, Richard M., editor, *The Horizon Book of the Arts of China*, American Heritage Publishing Company, Inc., New York, 1969.

Maddox, Brenda, *Reading the Rocks: How Victorian Geologists Discovered the Secret of Life*, Bloomsbury, New York, 2017.

Naish, Darren, August 4, 2012, https://blogs.scientificamerican.com/tetrapod-zoology/the-19th-century-discovery-of-dinosaurs/.

Peirce, C.S., "Illustrations on the Logic of Science II–VI: How to Make Ideas Clear," and "The Doctrine of Chances," Deduction, Induction and Hypothesis," *Popular Science Monthly*, vol. 12, January–August 1878.

Pennant, Thomas, *Synopsis of Quadrupeds*, Chester, Printed by J. Monk, from the collection of the Smithsonian Institution National Museum, Biodiversity Heritage Library, 1771.

Rudwick, Martin J.S., *Georges Couvier, Fossil Bones, and Geological Catastrophes*, University of Chicago Press, Chicago, 1997.

Thackray, John C., editor, *To See the Fellows Fight: Eyewitness accounts of meetings of the Geological Society of London and its Club, 1822–1868,* the British Society for the History of Science, London, 1999.

Torrens, Hugh, "Mary Anning (1799–1847) of Lyme; the 'greatest fossilist the world ever knew,'" *British Journal of the History of Science*, vol. 28, 1995, pp. 257–284.

Warren, Leonard, *Joseph Leidy: The Last Man Who Knew Everything*, Yale University Press, 1998.

Werner, E.T.C., *The Project Gutenberg eBook of Myths and Legends of China*, 2005, Project Gutenberg, www.gutenberg.org.

WGBH/pbs, *American Experience: Dinosaur Wars*, www.pbs.org/wgbh/americanexperience/films/dinosaur/.

Zhiming, Dong, and Milner, A.C., *Dinosaurs from China*, London Natural History Museum, Museum Publications, London, 1988.

GEOLOGIC LAWS AND PROCESSES

Baker, Victor R., "Catastrophism and uniformitarianism: logical roots and current relevance in geology," in *Lyell: The Past is the Key to the Present*, (Geological Society Special Publication), Geological Society, London, January 1998.

Blakemore, Erin, "Seeing Is Believing: How Marie Tharpe Changed Geology Forever," *Smithsonian Magazine*, August 30, 2016.

Buckland, William, *Geology and Minerology Considered with Reference to Natural Theology*, vols. 1–3, Pickering, London, 1836.

Cannon, W.F., "The Uniformitarianism-Catastrophism Debate," *Isis: A Journal of the History of Science Society*, vol. 51, no.1, March 1960.

Carroll, S.E., et al, editors, *Catalogue of Diatoms*, Book 7, the Elliss and Messina Catalogues of Micropaleontology, Micropaleontology Press, the American Museum of Natural History, New York, 1990.

Geological Society of London, www.geolsoc.org.uk/Plate-Tectonics/Chap1-Pioneers-of-Plate-Tectonics/Alfred-Wegener/Jigsaw-Fit.

Geological Society of London, London, Special Publications, 1998, 143, 171–182.

Gurer, Derya, "100 Years of Marie Tharp: the woman who mapped the ocean floor and laid the foundations of modern geology," *European Geosciences Union* blog, https://blogs.egu.eu/divisions/ts/2020/07/30/100-years-of-marie-tharp-the-woman-who-mapped-the-ocean-floor-and-laid-the-foundations-of-modern-geology/.

Hall, Stephen S., "The Contrary Map Maker," *New York Times Magazine*, December 31, 2006.

Lowry, William, and Alvarez, Walter, "One hundred million years of geomagnetic polarity history," *Geology*, vol. 9, no. 9, 1981, the Geological Society of America, 1981.

Lyell, Sir Charles, *Principles of Geology*, D. Appleton and Co., New York, 1838.

NASA, Earth Observatory, https://earthobservatory.nasa.gov/features/Steno.

Rudwick, Martin J.S., *Worlds Before Adam: The Reconstruction of Geohistory in the Age of Reform*, University of Chicago Press, Chicago, 2008.

Wetmore, Karen, "Foram Facts—An introduction to Foraminifera," https://ucmp.berkeley.edu/fosrec/Wetmore.html

ALVAREZ FAMILY HISTORY

Alvarez, Luis W., *Alvarez: Adventures of a Physicist,* Basic Books, New York, 1987.

Alvarez, Walter, *T. Rex and the Crater of Doom,* Princeton University Press, Princeton, NJ, 1997.

Alvarez, Walter and Milly, Email, August 13, 2020.

Alvarez, Walter C., *How to Help Your Doctor Help You,* Dell Publishing Company, Inc., New York, 1955.

Buffington, Andrew, and Sally, Interview, at their home in La Jolla, July 20, 2019.

Messinger, Gary, "Bay Area Kid: Growing Up on the West Coast in the 1950s," 2012, used by permission of the author.

Messinger, Gary, *California Remembered* blog, https://californiaremembered.wordpress.com/2015/01/29/walter-backk-then/.

Messinger, Gary, Phone Interview, June 18, 2020.

Nobel Prize, www.nobelprize.org/prizes/physics/1968/summary.

Sullivan, Walter, "Luis Alvarez, Nobel Physicist Who Explored Atom, Dies at 77," *New York Times*, September 2, 1988.

Trower, W. Peter, ed., *Discovering Alvarez: Selected Works of Luis W. Alvarez, with Commentary by His Students and Colleagues,* the University of Chicago Press, Chicago, 1987.

Weintraub, Pamela, and Gravenor, Misha, "The Man Who Discovered What Killed the Dinosaurs," *Discover,* October 26, 2009.

GEOLOGIC TIME

Boltwood, Bertram Borden, Papers, Yale University Archives, https://archives.yale.edu/repositories/12/resources/3799/collection_organization.

Buckland, William, *Geology and Minerology Considered with Reference to Natural Theology,* vols. 1–3, Pickering, London, 1836.

Cadbury, Deborah, *The Dinosaur Hunters,* Fourth Estate, London, 2000.

Linda Hall Library website, Scientist of the Day: www.lindahall.org/about/news/scientist-of-the-day/bertram-boltwood.

Linda Hall Library website, Scientist of the Day: www.lindahall.org/about/news/scientist-of-the-day/james-hutton/.

Lyell, Sir Charles, *Principles of Geology*, D. Appleton and Co., New York, 1838.

Maddox, Brenda, *Reading the Rocks: How Victorian Geologists Discovered the Secret of Life*, Bloomsbury, New York, 2017.

Winchester, Simon, *The Map That Changed the World: William Smith and the Birth of Modern Geology*, Harper Perennial, New York, 2009.

COSMIC RAYS AND IONIZING RADIATION

Buffington, Andrew, and Sally, Interview, July 20, 2019, at their home in La Jolla.

CERN, the European Organization for Nuclear Research, https://timeline.web.cern.ch/victor-hess-discovers-cosmic-rays.

Nobel Prize, www.nobelprize.org/prizes/physics/1936/hess/biographical/.

Tawa, Sebastien, PhD, Email, September 3, 2020.

Trower, W. Peter, ed., *Discovering Alvarez: Selected Works of Luis W. Alvarez, with Commentary by His Students and Colleagues*, the University of Chicago Press, Chicago, 1987.

ASTROGEOLOGY, STOKES'S LAW, AND CATASTROPHISM

Alvarez, L.W., Alvarez, W., Asaro, F., and Michel, H.V., "Extraterrestrial cause for the Cretaceous-Tertiary extinction: Experimental results and theoretical interpretation," *Science*, vol. 208, 1980, pp. 1095–1108.

Baker, Victor R., "J. Harlen Bretz (1882–1981), Outrageous Geological Hypothesizer," *Geology Today*, May 2022.

Bretz, J. Harlen, plaque on monument at Dry Falls State Park, Washington.

Fowler, Michael, Lectures on Fluids, Physics 152, University of Virginia, https://galileo.phys.virginia.edu/classes/152.mf1i.spring02/Stokes_Law.htm.

Russell, Dale, and Tucker, Wallace, "Supernovae and the Extinction of the Dinosaurs," *Nature*, vol. 229, February 19, 1971.

SEDIMENTOLOGY AND TSUNAMIS

Alaska Earthquake Center, https://earthquake.alaska.edu/60-years-ago-1958-earthquake-and-lituya-bay-megatsunami.

BBC Nature, *Megatsunami: Evidence of Destruction*, BBC Studios, https://youtu.be/2uCZjqoRLjc?si=WfRMsEkmCpbNAJvZ.

BBC Nature, *Megatsunami: Lituya Bay Survivors,* Interview with Sonny and Howard Ulrich, www.dailymotion.com/video/xhqagp.

BBC2 Science, *Megatsunami: Wave of Destruction,* transcript, October 12, 2000, www.bbc.co.uk/science/horizon/2000/mega_tsunami_transcript.shtml.

Bourgeois, Joanne, "Boundaries: A Strategraphic and sedimentologic perspective," *Geological Society of America,* Special Paper 247, 1990.

Bourgeois, Joanne, et al., "Sedimentological effects of tsunamis, with particular reference to impact-generated and volcanogenic waves," Abstracts Presented to the Topical Conference on Global Catastrophes in Earth History: An Interdisciplinary Conference on Impacts, Volcanism, and Mass Mortality, Snowbird conference, 1988.

Bourgeois, Joanne, Phone Interview, September 14, 2019.

Hansen, Thor (Western Washington University), "K-T Highlights: Cretaceous-Tertiary Boundary in Texas," PowerPoint/lecture, University of Washington, 1986.

Hildebrand, A.R., et al., "Chicxulub crater: A possible Cretaceous/Tertiary boundary impact crater on the Yucatán Peninsula, Mexico," *Geology,* vol. 19, 1991, pp. 867–71.

Pope, K., Ocampo, A., and Duller, C., Mexican site for K/T impact crater? *Nature* (Scientific Correspondence), 351, 105 (1991).

Schwing, Patrick, "Little Critters That tell a BIG Story: Benthic Foraminifera and the Gulf Oil Spill," *Smithsonian Institution,* June 2014, https://ocean.si.edu/contributors/patrick-schwing

GEOLOGIC EVIDENCE OF AN ASTEROID IMPACT

Bohor, Bruce, F., "Shocked Quartz and More: Impact Signatures in K-T Boundary Clays and Claystones," *Abstracts Presented to the Topical Conference on Global Catastrophes in Earth History: An Interdisciplinary Conference on Impacts, Volcanism, and Mass Mortality,* Lunar and Planetary Institute and the National Academy of Sciences, Snowbird, UT, October 23, 1988, p. 17.

Claeys, P., Kiessling, W., and Alvarez, W., "Distribution of Chicxulub ejecta at the Cretaceous-Tertiary Boundary," in Koeberl, C., and MacLeod, K.G., eds., *Catastrophic Events and Mass Extinctions: Impacts and Beyond,* Geological Society of America Special Paper, Boulder, CO, 2002, pp. 241–57.

Penfield, Glen, Phone Interview, October 26, 2024.

Schulte, P., et al., "The Chicxulub asteroid impact and mass extinction at the Cretaceous-Paleogene boundary," *Science,* vol. 327, 2010, pp. 1214–18.

Schuraytz, B.C., et al., "Iridium metal in Chicxulub impact melt: Forensic chemistry on the K-T smoking gun," *Science,* vol. 271, 1996, pp. 1573–76.

Simonson, B.M., and Glass, B.P., "Spherule layers—Records of ancient impacts," *Annual Review of Earth and Planetary Sciences*, vol. 32, 2004, pp. 329–61.

Smit, J., "The global stratigraphy of the Cretaceous-Tertiary boundary impact ejecta," *Annual Review of Earth and Planetary Sciences*, vol. 27, 1999, pp. 75–113.

Smit, J., Zoom Interview, July 11, 2020.

Smit, J., and Hertogen, J., "An extraterrestrial event at the Cretaceous-Tertiary boundary," *Nature*, vol. 285, 1980, pp. 198–200.

Smit, J., and Romein, A.J.T., "A sequence of events across the Cretaceous-Tertiary boundary," *Earth and Planetary Science Letters*, vol. 74, 1985, pp. 155–70.

Wilford, John Noble, "For Dinosaur Extinction Theory, a 'Smoking Gun,'" *New York Times*, February 7, 1991.

ALTERNATIVE HYPOTHESES, DECCAN VOLCANISM, AND CONTROVERSIES

Benton, M.J., "Scientific Methodologies in Collision: The History of the Study of the Extinction of the Dinosaurs," *Evolutionary Biology*, 1990.

Bosker, Bianca, "The Nastiest Feud in Science," *The Atlantic*, September 2018.

Brown, Malcolm, "The Debate over Dinosaur Extinctions . . ." *New York Times*, January 19, 1988.

Keller, Gerta, "Extinction, survivorship and evolution of planktic foraminifera across the Cretaceous/Tertiary boundary at El Kef, Tunisia," *Marine Micropaleontology*, January 10, 1988.

Keller, Gerta, Phone Interview, October 25, 2019.

MacLeod, Norman, and Keller, Gerta, *Cretaceous-Tertiary Mass Extinctions: Biotic and Environmental Changes*, W.W. Norton & Company, New York, 1996.

Officer, Charles, and Page, Jake, *The Great Dinosaur Extinction Controversy*, Helix Books, Addison-Wesley Publishing Company, Inc., Menlo Park, CA, 1996.

Punekar, Jahnavi, Mateo, Paula, and Keller, Gerta, "Effects of Deccan Volcanism on paleoenvironment and planktic foraminifera: a global survey," *Geological Society of America Special Papers*, August 21, 2014.

Renne, P.R., et al., "State shift in Deccan volcanism at the Cretaceous-Paleogene boundary, possibly induced by impact," *Science*, vol. 350, 2015, pp. 76, 78.

Richards, M.A., et al., "Triggering of the largest Deccan eruptions by the Chicxulub impact," *Geological Society of America Bulletin*, vol. 127, no. 11–12, pp. 1507–20.

Schoene, B., et al., "U-Pb geochronology of the Deccan Traps and relation to the end-Cretaceous mass extinction," *Science*, vol. 347, 2015, pp. 182–84.

CURRENT RESEARCH ON THE CRETACEOUS EXTINCTION

Even as this book goes to press, scientists continue to explore the asteroid impact and Cretaceous extinction. They are researching everything from the season of the impact event and its immediate aftermath, to the origin of the asteroid in space, to which chemicals in the atmosphere influenced the severity of the impact winter and how the extinction unfolded. The following articles, scientific papers, and scientists helped me to understand where research is heading, and the questions scientists will be asking in the future.

DePalma, Robert, Smit, Jan, et al., "A seismically induced onshore surge deposit at the KPg boundary, North Dakota," 2019, https://doi.org/10.1073/pnas.1817407116.

During, Melanie A.D., Smit, J., et al., "The Mesozoic terminated in boreal spring," *Nature*, vol. 603, March 3, 2022.

Henehan, Michael J., et al. (incl. Hull), "Rapid ocean acidification and protracted Earth system recovery followed the end-Cretaceous Chicxulub impact," *Procedures of the National Academy of Sciences*, vol. 116, no. 45, November 5, 2019, pp. 22500–04.

Hull, Pincelli, "Pincelli Hull Explains What Killed Off the Dinosaurs," Interview with *Quanta* online magazine, www.quantamagazine.org/videos/pincelli-hull-explains-what-killed-off-the-dinosaurs/.

JOIDES Resolution, *How Science Works*, Video, Pincelli Hull interview starting at minute 5:28, www.youtube.com/watch?v=i9tsdAQBcfM.

Junium, Christopher K., Zerkle, Aubrey L., et al., "Massive perturbations to atmospheric sulfur in the aftermath of the Chicxulub impact," *PNAS*, vol. 119, no. 14, March 21, 2022.

Kortsha, Monica, "Asteroid Dust Found in Crater Closes Case of Dinosaur Extinction," *UT News*, February 24, 2021, www.news.utexas.edu/2021/02/24/asteroid-dust-found-in-crater-closes-case-of-dinosaur-extinction/.

Levy, Renee, "The Impact of Extinction: New research published in PNAS answers a lingering question about the source of atmospheric sulfur leading to the extinction of the dinosaurs," https://artsandsciences.syracuse.edu/news-all/news-from-2022/the-impact-of-extinction/.

Lowery, Christopher, Zoom Interview, July 7, 2020.

Nesvorny, David, Bottke, William F., and Marchi, Simone, "Dark primitive asteroids account for large share of K/Pg-scale impacts on Earth," *Icarus*, vol. 368, November 2021.

Pennisi, Elizabeth, "How Life Blossomed After the Dinosaurs Died," *Science*, October 24, 2019.

Redd, Nola Taylor, "After the Dinosaur-Killing Impact, Soot Played a Remarkable Role in Extinction," *Smithsonian Magazine*, April 27, 2020.

Science Desk, "Where did the dinosaur-killing Chicxulub asteroid come from?," *The Indian Express*, www.indianexpress.com/article/technology/science/dinosaur-killing-chicxulub-asteroid-origin.

Sibert, Elizabeth, Email, July 29, 2020.

Smit, J., Interview in Person, Amsterdam, July 21, 2023.

Southwest Research Institute, "SwRI Team Zeroes in on Source of the Impactor That Wiped Out the Dinosaurs," July 28, 2021, www.swri.org/press-release/swri-team-zeroes-source-wiped-out-dinosaurs.

Tabor, Clay R., et al., "Causes and Climatic Consequences of the Impact Winter at the Cretaceous-Paleogene Boundary," *Geophysical Research Letters*, January 15, 2020.

Zolfagharifard, Ellie, "How 2011 Japanese earthquake created freak 5ft waves that terrified locals on the other side of the world—in Norway," *Daily Mail*, London, August 19, 2013.

DIVERSITY IN US GEOSCIENCES

And finally, in interviewing female scientists—especially during the pandemic—and reading their scientific papers and accounts of their lives, I became painfully aware that many of them faced, and continue to face, gender-based obstacles in their path to success. I met women who, like Mary Anning, did not receive proper credit for their discoveries, which were often stolen and published under men's names. I met women who faced subtle but real hostility and disregard in the workplace. And I met PhDs who worked the "second shift" daily, taking on the lion's share of parenting and household work while continuing their important scientific work, especially during the pandemic. Their stories pushed me to include a sidebar about diversity in the geosciences in the hope that one day these barriers will fall and we will have true racial and gender equality in the sciences.

Bernard, Rachel E., and Cooperdock, Emily H.G., "No progress on diversity in 40 years," *Nature Geoscience*, vol. 11, May 2018.

Dutt, Kuheli, "Race and racism in the geosciences," *Nature Geoscience*, vol. 13, January 2020.

ACKNOWLEDGMENTS

I would like to thank all the scientists mentioned below for their help in researching and writing this book. I am indebted to them for inspiring me and my students to dig deeper into the world, and to try a little harder. I admire them for their dedication to doing the work of science—in the field, under unimaginably difficult conditions, in the lab, and in society at large. I appreciate them for their commitment to communicating the wonder of science to the world. And I value their struggles to make the science world a more equitable place. Everyone has a right to be on the cutting edge of discovery, and we all have a responsibility to open the doors of science wider.

Thank you especially to Dr. Richard Norris, Scripps Institution of Oceanography, for reading the entire manuscript and suggesting a whole new sixth chapter, and then reading that chapter through a second time; for suggesting young researchers for me to contact; and for digging up articles for me to read.

Thank you, Alexandra Hangsterfer, for introducing me and my students to the SIO Deep Sea Drilling Project Geology Locker, a repository of drill cores from around the world, and to Kalle Palmer, high school biology teacher and diving instructor extraordinaire, for introducing me to Alexandra.

Special thanks to Jan Smit, PhD, paleontologist, VU University of Amsterdam, for seeing this project through to the end, going out of his way to offer long and very entertaining interviews, taking me on a field trip to see the K/Pg boundary, sharing photos, clarifying information, articles, good cheer and goodwill, and introducing me to young scientists as I pieced together this Cretaceous puzzle.

The following scientists sent me articles to read, spent hours with me over the phone and on Zoom interviews, and/or volunteered to read chapters of the manuscript and offered corrections and suggestions to make sure I got the science right. Any mistakes remaining rest squarely on my own shoulders.

- Jody Bourgeois, PhD, sedimentologist, University of Washington
- Andy Buffington, PhD, astrophysicist emeritus, University of California, San Diego, and his wife, Sally
- Melanie During, PhD candidate, paleontology
- Pincelli Hull, PhD, geologist, Yale University
- Gerta Keller, PhD, geologist, Princeton University
- Christopher Lowery, PhD, geophysicist, University of Texas Institute for Geophysics
- Glen Penfield, geologist, retired
- Sebastien Tawa, PhD, astrophysicist

Thank you to Gary Messinger, PhD, for sharing his beautifully written as-yet-unpublished memoir about growing up in the Berkeley Hills with his childhood neighbor Walter Alvarez, and for chatting with me about his experience.

And finally, thank you to Walter and Milly Alvarez, who very kindly took time to read parts of the completed manuscript and offer suggestions for changes.

Thank you from the bottom of my heart to Stefanie Sanchez von Borstel, my agent at Full Circle Literary, LLC, for suggesting I submit a proposal for this book at a time when I was returning to full-time teaching and could barely think of writing at all. Thanks for sticking with me and keeping the project moving forward.

My deepest thanks to editors Cheryl Klein, Adam Rau, and Stacy Whitman, managing editor Melissa Kavonic, art director Neil Swaab, copyeditor Jill Amack, proofreader Chandra Wohleber, all the staff at Lee & Low, and glorious illustrator Theo Nicole Lorenz, who have stuck with this project through school-year demands, graduations, surgeries, job changes, the birth of a baby, the deaths of a parent and a spouse, a (tiny) tsunami, and, of course, a pandemic—everything short of an asteroid. Thanks for believing that science—and this particular story—matters, and that I could tell it even when I wasn't so sure myself.

Special thanks go to Elijah Bonde, Gil Brady, and all the staff, students, and board members of Nativity Prep Academy, Carver Elementary, Madison High School, and UCSD Extended Studies for supporting me in my teaching work so I can continue writing. Thank you especially to my students, who inspire and delight me with their curiosity and enthusiasm.

And finally, thank you to my family, especially my husband, Chris, who supported me with time and love, and died while this book was in production. Thank you to my children, Ronan and Eddie, who are always willing to geek out about discoveries. Thank you to my siblings, Jim, Julie, and Laurie, who buoy me up each week. And finally, I write in honor of my parents—Bill, who passed away as I was working on this project, and who was always willing to be a fanboy at book signings, and my mother, Norma, who typed my stories when I was a little girl so I would feel like a real writer by seeing my work in print. I do, Mom, thanks to you.

ENDNOTES

GRAPHIC NONFICTION ONE
Alamosaurus, etc., Smit, Email, July 11, 2020.
Edmontosaurii, Smit, Email, July 11, 2020.

CHAPTER ONE
"GeolSoc," Maddox, *Reading the Rocks*, 48.
"Toast of mice," Cadbury, *The Dinosaur Hunters*, 61.
"As a working class woman," Maddox, *Reading*, 46.
"Notice on the Discovery," Cadbury, 106, Thackray, 4, Maddox, 54.
"To the head of a lizard," Cadbury, *The Dinosaur Hunters*, 107-108.
"Fish and shellfish," Cadbury, *The Dinosaur Hunters*, 109.
"The first scientific description," Maddox, *Reading*, 74.
"With two monsters," Thackray, *To See*, 4.
"Near to reptile," Maddox, *Reading*, 53.
"According to her account," Cadbury, *The Dinosaur Hunters*, 231.
"Dragon bones," Dong Zhiming, *Dinosaurs from China*, 9.
"Thigh bones of a man," Cadbury, *The Dinosaur Hunters*, 63.
"As yet the living animal," Pennant, *Synopsis*, 92.
"Natural theology," Buckland/*Geology*, 1.
"The 'Creator'" had placed," Benton/Scientific Methodologies, 3.
"Science first begins," Peirce/*Illustrations* III, 1.
"The processes of investigation," Peirce/*Illustrations* II, 1.
"Science practices," https://www.nextgenscience.org.
"Bone Wars," WGBH/ pbs/dinosaur.

"Between them," www.pbs.org/wgbh/americanexperience/films/dinosaur/.

"If the coelacanth," Colbert, 1965, 250.

"Scientists developed both," Benton, 5–9.

GRAPHIC NONFICTION TWO

Nicholaus Steno, NASA, July 2004.

Harry Hess, Geological Society, www.geolsoc.org.uk/Plate-Tectonics/Chap1-Pioneers-of-Plate-Tectonics/Harry-Hess.

CHAPTER TWO

"Do you want to go," Messinger, *California Remembered* blog, accessed June 2020, Interview, June 18, 2020.

Luis Fernandez Alvarez, Alvarez, Luis, *Adventures*, 9.

Walter Clement Alvarez, *Adventures*, 10–11.

"America's Family Doctor," Alvarez, Luis, *Adventures*, 13.

"All the scientists," Alvarez, Luis, *Adventures*, 14.

Geraldine Smithwick, Alvarez, Luis, *Adventures*, 37.

"Dad and I talked," Alvarez, Luis, *Adventures*, 58.

"If you have watches," Messinger, Interview, June 18, 2020.

"Inside microscopic particles spin," Messinger, Interview, June 18, 2020.

"Get out of the way," Messinger, Interview, June 18, 2020.

"Now let me show," Messinger, Interview, June 18, 2020.

"This is how," Messinger, Interview, June 18, 2020.

"Do you want," "I built this," Messinger, *California Remembered* blog, accessed June 2020.

"Walter was born," Alvarez, Luis, *Adventures*, 85.

Bob Bacher, Alvarez, Luis, *Adventures*, 123.

"I believe," Alvarez, Luis, *Adventures*, 151–152.

"Rock hammer," Alvarez, Walter, *T-Rex*, 60.

"I like being out," Weintraub and Graenor, "The Man," *Discover*, 25.

"Map-maker," Hall, www.nytimes.com/2006/12/31/magazine/31Tharp.t.html.

"Magnetic polarity," Alvarez, Walter, *T-Rex*, 36.

"Far-flung places," Alvarez, Walter, *T-Rex*, 46.

"A former girl scout," Alvarez, Walter and Milly, Email, August 2020.

"Foraminifera," Wetmore, https://ucmp.berkeley.edu, 1.
"The question of," Alvarez, Walter, *T-Rex*, 42.
"Occasionally there is a question," Alvarez, Walter, *T-Rex*, 42.

GRAPHIC NONFICTION THREE

"Why did God," Maddox, *Reading*, 10.
"The ruins of an older world," Maddox, *Reading*, 10.
"No vestige of a beginning," Maddox, *Reading*, 11.
"He took a boat," Maddox, *Reading*, 14.
"Principles of Geology," Maddox, *Reading*, 150.
"Greywacke," Cadbury, *The Dinosaur Hunters*, 21–22.
"A former soldier," Maddox, *Reading*, 112–116.
Bertram Boltwood, www.lindahall.org/about/news/scientist-of-the-day/bertram-boltwood/.
"Geologic time scale," Maddox, *Reading*, 203.

CHAPTER THREE

Victor Hess, CERN, https://timeline.web.cern.ch/victor-hess-discovers-cosmic-rays-0.
"Mexico City," Alvarez, Luis, *Adventures*, 26.
"They valued," Alvarez, Luis, *Adventures*, 214.
"For his decisive contributions," www.nobelprize.org/prizes/physics/1968/summary.
"Over the next six months," Alvarez, Luis, *Adventures*, 216.
"Since most physicists," Alvarez, Luis, *Adventures*, 47.
"If Luis thought," Buffington, Interview, July 20, 2019.
"Inside an Egyptian pyramid," Alvarez, Luis, "Using Cosmic Rays . . ." in Trower, 181.
"Luis was a 'want to know' person," Buffington, Interview, July 20, 2019.
"The only way I could learn," Alvarez, Luis, *Adventures*, 46.
"Beryllium-10," Alvarez, Walter, *T-Rex*, 62.
"Iridium," Alvarez, Luis, *Adventures*, 253.
"As the reactor fired," Alvarez, Luis, *Adventures*, 253.
"Walter brought twelve samples," Alvarez, Walter, *T-Rex*, 68.
"Closer to 9 ppb," Alvarez, Walter, *T-Rex*, 69.

GRAPHIC NONFICTION FOUR

Tiny craters, Alvarez, Walter, *T-Rex*, 51.

Scablands, Alvarez, Walter, *T-Rex*, 51.

"Great Spokane Flood," Video, www.pbs.org/video/ksps-documentaries-sculpted-by-floods-the-northwests-ice-age-legacy/.

"Ideas without precedent," Bretz, plaque at Dry Falls Visitor Center, Coulee City, WA.

CHAPTER FOUR

"Fish clay," Alvarez, Walter, *T-Rex*. 70.

"Supernova," Russell and Tucker, *Nature*, vol. 229, 553–554.

"You're barking up the wrong tree," Buffington, Interview, July 20, 2019.

"One in a million," Trower, *Discovering Alvarez*, 241.

"Do it all over again," Alvarez, Walter, *T-Rex*, 74.

"Chocolate chip cookies and strawberry ice cream," Trower, *Discovering Alvarez*, 241.

"Wave," Alvarez, Luis, *Adventures,* 255, Alvarez, Walter, *T-Rex*, 77.

"Ideas," Trower, *Discovering Alvarez*, 241.

"Year without summer," University Center for Atmospheric Research: https://scied.ucar.edu/learning-zone/how-climate-works/mount-tambora-and-year-without-summer.

"The smoke would obscure the sun," Trower, *Discovering Alvarez*, 242.

Stokes's Law, Fowler, "Lectures on Fluids," https://galileo.phys.virginia.edu/classes/152.mf1i.spring02/Stokes_Law.htm.

"We've got the answer," Alvarez, Walter, *T-Rex*, 77.

"Extraterrestrial Cause," Alvarez, Alvarez, et al., *Science,* June 1980.

"Stamp collectors," Brown, www.nytimes.com/1988/01/19/science/the-debate-over-dinosaur-extinctions-takes-an-unusually-rancorous-turn.html.

"The Nastiest Feud in Science," Bosker, *The Atlantic,* September 2018, 1.

GRAPHIC NONFICTION FIVE

"Facing the wave," *Megatsunami: Lituya Bay Survivors*, BBC Nature.

"Debris of giant logs," Alaska Earthquake Center, https://earthquake.alaska.edu/60-years-ago-1958-earthquake-and-lituya-bay-megatsunami.

"The Ulriches," *Megatsunami: Lituya Bay Survivors*, BBC Nature, https://youtu.be/2uCZjqoRLjc?si=k7WodVevAeFUoOuq

"Evidence of past tsunamis," BBC2 Science, transcript, October 12, 2000.

"K-T boundary," Hansen, PowerPoint, 1986.

"Tsunami!" Bourgeois, Interview, September 14, 2019.

CHAPTER FIVE

"Leeches," Bourgeois, Interview, September 14, 2019.

"One of the best," Smit, Interview, July 11, 2020.

"In North America," Officer, *The Great . . . Controversy*, 65–66.

"Capstone event," Officer, *The Great . . . Controversy*, 45.

"Pulses of flowing lava," Keller, "Extinction," 1988, *Marine Micropaleontology*, 239–263.

"Walter's colleague who had studied more," Alvarez, Walter, *T-Rex*, 108.

"He concluded that," Bohor, "Shocked Quartz," *Abstracts*, October 23, 1988, 17.

"This may be the first," Smit and Romein, "Sequence of Events," *Earth and Planetary Science Letters*, 160.

"Tell-tale trace," Smit, Email, July 2020.

"Alvarez seemed to care," Sullivan, "Luis Alvarez," *New York Times*, September 3, 1988, 1.

"magnetometer," Penfield interview, October 26, 2024.

"Drilling Leg 77," Alvarez, Walter, *T-Rex*, 111.

"Cretaceous," Smit, interview, July 11, 2020.

"Penfield sent Walter," Penfield interview, October 26, 2024.

"Smoking gun," Wilford, "For Dinosaur Extinction," *New York Times*, February 7, 1991.

Chicxulub sequence graphic, Smit, Email, August 2020.

CHAPTER SIX

"If you take the JR," JOIDES Resolution, "How Science Works," Video.

"If you come into the core lab," JOIDES Resolution, Video.

"Telling stories about past," Sokol, "Pincelli Hull Explains," *Quanta*, March 25, 2020.

"Sending many species," Keller, "Extinction," *Marine Micropaleontology*, 1988.

"Her work suggested," Punekar, Mateo, and Keller, "Effects," *Geological Society of America Special Papers*, August 21, 2014, 91–105.

Jan Smit "raised a finger," Smit, Email, August 2020.

"The nastiest feud in science," Bosker, *The Atlantic*, September 2018.

"Why do certain species," Keller, Interview, phone, October 25, 2019.

"The impact happened," Keller, Interview, phone, October 25, 2019.

"We collect samples," Keller, Interview, phone, October 25, 2019.

"Survivor species," Punekar et al., August 21, 2014, 91–105.

"Some species could survive," Punekar et al., August 21, 2014, 91–105.

"It turns out," Sokol, *Quanta*, March 25, 2020.

"In one study," Henehan et al., 3.

"Allowed us to say," Sokol, "Pincelli Hull,"*Quanta*, March 25, 2020.

"Increased diversity," Bernard and Cooperdock.

"Transition zone," Lowery, Interview, July 7, 2020.

"What we discovered," Lowery, Interview, *UT News*, 2021.

"Two critical [questions]," Southwest Research Institute, July 28, 2021, www.swri.org/press-release/swri-team-zeroes-source-wiped-out-dinosaurs.

"Decided to look," Southwest Research Institute, July 28, 2021, www.swri.org/press-release/swri-team-zeroes-source-wiped-out-dinosaurs.

"Asteroid belt," Nesvorny, "Dark, Primitive Asteroids," *Icarus*, November 2021.

"Logjams," Preston, "The Day," *New Yorker*, April 8, 2019, 54.

"Since then the evidence," Sanders, "66-Million," *Berkeley News*, March 29, 2019.

"He found," Smit, Interview, July 11, 2020.

"There are eggs," Smit, Interview, July 11, 2020.

"I assure you . . . this was the first time," Smit, Interview, July 11, 2020.

"Norway seiche waves," Zolfagharifard, "How 2011," *Daily Mail*, August 19, 2013.

"The day the dinosaurs died," Preston, *New Yorker*, April 8, 2019, 52–65.

"A Seismically Induced," DePalma et al., *PNAS*, 2019, https://doi.org/10.1073/pnas.1817407116.

"The Mesozoic Terminated in boreal spring," During, *Nature*, March 3, 2022.

"The Impact of Extinction," Levy, "The Impact," https://thecollege.syr.edu/news-all/news-from-2022/the-impact-of-extinction.

"Right now, we are," Alvarez, Walter, *T-Rex*, 143–144.

"Had delighted in the effort," Alvarez, Walter, *T-Rex*, 143–144.

INDEX

A Group (University of California, Berkeley), 75–77
Academy of Natural Sciences (Philadelphia), 28, 29
Acid rain, 158, 171–173
Acidification, ocean, 167
Ages (in time scale), 74
Alaska, 111–122
Allosaurus, 29
Alvarez, Geraldine. *See* Smithwick, Geraldine
Alvarez, Jan, 75, 77–79
Alvarez, Jean, 46, 51, 78
Alvarez, Luis Fernandez, 43–44
Alvarez, Luis Walter
  on authority, 45, 47
  childhood of, 45–46
  death of, 138
  debating other scientists, 108
  on formation of K-T boundary, 84
  A Group of, at University of California, 75–77
  impact hypothesis by, 103–107
  lab of, at University of California, 46, 47–50
  at Los Alamos, 50, 51
  Nobel Prize of, 77–80
  photos of, 44, 76, 77, 78, 107, 185
  radiometric dating idea of, 83
  studies of, 46
  wives of (*See* Alvarez, Jan; Smithwick, Geraldine)
  work on Egyptian pyramids, 81
  work with Alvarez (Walter), 82–87, 95–109, 184–186
Alvarez, Milly. *See* Millner, Mildred "Milly"
Alvarez, Walter
  in Apennine Mountains, 60–65, 81
  on beryllium-10, 83
  childhood of, 43, 47–52
  in debate with other scientists, 109
  drill core samples tested by, 146–147
  on formation of K-T boundary, 84
  on impact hypothesis, 104–107, 146–147
  in impact site search, 133, 134, 139, 142–144, 146–147
  on iridium, 107, 129
  photos of, 44, 107, 179, 185
  at Stevns Klint site, 95–97
  studies of, 52, 58
  on supernova theory, 99–100
  at Tanis fossil site, 179–182
  travels and work of, 59
  wife of (*See* Millner, Mildred "Milly")

work with Alvarez (Luis Walter), 82–87, 95–109, 184–186
work with Lowrie (Bill), 60–65
Alvarez, Walter Clement, 44–46
Alvarez: Adventures of a Physicist (Alvarez), 51
Alvarez hypothesis. *See* Impact hypothesis
Alvin (submersible vehicle), 55
American measurements, 2
American Museum of Natural History (New York City), 30
Amphicoelias, 28
Analyzing, as scientific practice, 2
Andesite, 141
Anning, Mary, 16, 17, 19
Anomaly, gravity, 140–141, 190
Apennine Mountains, 60–65, 81
Apollo 11 mission, 89
Archean Eon, 74
Arguments
  scientific, 80
  using evidence for, 2
Arizona, 132–133
Arroyo el Mimbral, Mexico, 143–146
Asaro, Frank, 85–87, 99, 101, 105, 107
Ash, 158
Ashmolean Museum, 21
Asking questions, 1, 27
Asteroid(s)
  as cause of K-T extinction, 103–109, 149–158, 173
  defined, 188
  description of, 102
  discussions about, 101–102
  source of, 175–176
Astrogeology, 92, 188
Astronomer, 188
Astrophysics, 32, 76, 188
Atlantic Ocean floor, 40, 53
Atomic bomb, 50, 51
Atoms, 78–79. *See also* Subatomic particles

Bacher, Bob, 51
Bang, Inger, 95–96
Barringer Crater, 133
Basalt, 129
Beloc, Haiti, 136
Bernard, Eleanor, 172
Beryllium-10, 83
Biological diversity, 172
Bohor, Bruce, 131–132
Boltwood, Bertram, 73
Bone Wars, 28–30
Bottke, William, 176

Bourgeois, Jody (Joanne), 120–122, 123–125, 135–137, 184
Brazos River site, 123, 134–137, 139
Bretz, J. Harlen, 93–94
Buckland, Mary Morland, 19
Buckland, William, 15–18, 19, 24, 71
Buffington, Andrew, 77, 80, 83, 98–99

Calcium, 129
Cambrian Period, 72, 74
Carbon dioxide, 156, 167, 169
Carbon-12, 182
Carbon-13, 182
Carbonaceous chondrites, 176
Carbonic acid, 167
Carboniferous Period, 72
Carleton College, 52, 58
Cartographer, 53, 188
Catastrophism
  Bretz (J. Harlen) on, 93–94
  defined, 188
  development of concept of, 23–24
  vs. gradualism, 33
  vs. uniformitarianism, 25, 26, 33, 94, 107
Catastrophists, 24, 33, 188
Cenozoic Era, 74
Ceres (asteroid), 102
Challenging, as scientific practice, 2, 27
Chao, Edward, 92
Chickens, 27
Chicxulub crater, 141, 142, 145, 146, 173–176
China, fossils found in, 20
Church of England, 24
Claeys, Philippe, 145
Clark, William, 22–23
Classification, 18, 21–22, 188
Clemens, Bill, 108
Climate change, 31, 33
Cloud chamber, 48–49
Coelacanth, 30
Coesite, 92
Collaboration, 1
Colorado, 108, 109
Columbia University, 60, 136
Comets, 102
Communicating, as scientific practice, 2
Compass, 54
Conferences, scientific, 126, 137, 162, 169
Continental crust, 129, 130, 145, 146, 188
Continental drift, 36–39, 53, 188
Convection currents, 56, 188
Conybeare, William Daniel, 16–18
Cooperdock, Emily H. G., 172
Cope, Edward Drinker, 28–29

212

Cosmic dust, 84–87, 97, 175, 189
Cosmic ray detection experiments, 75, 76
Cosmic rays, 76, 77, 97, 189
Craters
 drilling, 173–176
 on Earth, 90–94
 infilling of, 159, 174–175
 on the moon, 89
 search for K-T boundary impact site, 131–137, 139–147
 in space, 90, 102
Cretaceous Period, 3–11, 35, 63–64, 72, 74, 149. *See also* K-T boundary
Crust
 continental, 129, 130, 145, 146, 188
 defined, 189
 discovery of, 53
 formation of, 53–55, 56
 oceanic, 129, 130, 134, 146, 191
 plate tectonics and, 53–57, 192
Crystals, 129–130
Cuvier, Georges, 23–24
Cyclotron, 48–49, 189

Dartmouth College, 127
Darwin, Charles, 26
Data gathering, 1, 27
Dating, radiometric, 73, 83, 146, 175
Deccan Traps (India), 127–128, 169, 173
Denmark, 95–97
DePalma, Robert, 176–184
Deposits/deposition, 69, 189
Devonian Period, 72, 74
Dinosaur extinction hypotheses, 31–32. *See also* Impact hypothesis; Volcanism theory
Dinosaur fossils
 classification of, 18, 21–22
 early discoveries, 17–19, 20–21
 found in China, 20
 found in England, 17–19, 21
 found in North America, 28–30
 public fascination with, 29–30
Dinosauria, 18, 20–22
Discourse on the Revolutionary Upheavals of the Earth (Cuvier), 23
Discovering Alvarez (Trower), 138
Diversity
 biological, 172
 in geosciences, 172
Doyle, Sir Arthur Conan, 30
Dragons, 20
Drill cores
 Chicxulub site, 173–176
 defined, 139–140, 189
 JOIDES Resolution, 165–167
 library of, 140
 Manson, Iowa, 133
 Ocean Drilling Project, 139–140
 PEMEX survey, 141–142, 146
Dryptosaurus aquilinguis, 28
During, Melanie A. D., 182
Dust
 cosmic, 84–87, 97, 175, 189
 lunar, 89
 Stokes's Law on, 105–106
Dutt, Kuheli, 17

Earth
 age of, 67–74
 craters on, 90–94
 formation of, 93–94
Earthquakes
 asteroid strike causing, 180
 map of, 41
 strata removed by, 82

tectonic plates and, 55, 57
tsunami caused by, 111–117
Egyptian pyramids, 81
Ejecta, 133–134, 144, 152, 189
Electromagnet, 48–49
Electroscope, 76, 189
England
 catastrophism in, 24, 25, 26
 fossils found in, 16–19, 21
 uniformitarianism in, 25, 26
Eocene Epoch, 74
Eons, 74
Epochs, 74
Eras, 74
Erosion, 82, 189
Ethnic minorities, in geosciences, 172
Evidence
 Darwin (Charles) presenting, 26
 in peer-review process, 63
 using for arguments, 2
Evolution through natural selection, 26, 189
Ewing, Maurice, 53
Extinction, 22–26. *See also* K-T boundary
 catastrophism on, 23–24, 33
 as concept, 22
 defined, 189
 evolution through natural selection and, 26
 gradualism on, 33
 mass, 98–99, 102, 126, 167–170, 183
 religious views on, 23, 24
 uniformitarianism on, 25, 26, 33
"Extraterrestrial Cause for the Cretaceous-Tertiary Extinction" (Alvarez, Alvarez, Michel, and Asaro), 107

Fairweather Fault, 111–117
Feldspar, 129, 130
Fermi, Enrico, 51
Ferns, 31, 160
Fiction, 30
Films, 30
Finches, 26
Fish fossils, 22, 178, 180, 181, 182
Flowering plants, 161
Foraminifera (forams), 61–65, 135–136, 143, 170–171, 189
Fossils
 classification of, 18, 21–22
 early discoveries, 16–19, 20–21
 formation of, 12
 found in China, 20
 found in England, 16–19, 21
 found in North America, 28–30
 at K-Pg site, 177–182
 presented to Geological Society, 16–18
 process of finding, 177
 in strata, 35, 71–72
 study of (*See* Paleontology)
Galapagos Islands, 26
Gamma rays, 86–87, 99, 189
Gamper, Martha, 142
Geiger counter, 46
Gender discrimination
 in geosciences, 172
 in paleontology, 16, 19
Geologic Time Scale, 74
Geological change, 31–32
Geological Society of America, 94, 184
Geological Society of London, 15–18, 19, 21, 25, 71, 72
Geology, 52–55, 189

Geosciences, diversity in, 172
Gertie the Dinosaur (film), 30
Glass, impact-melt, 145–147
Gradualism/gradualist, 33, 190
Graphite, 51
Gravity anomaly, 140–141, 190
Gravity field, 140, 141, 190
The Great Dinosaur Extinction Controversy (Officer), 127
Great Global Rift, 53
Great Spokane Flood, 93
Gregersen, Soren, 95–96
Greywacke, 72
Grieve, Richard, 132
Guajira Peninsula, 58, 59
Gulf of Mexico, 140–144, 153
Gulick, Sean, 175

Hadean Eon, 74
Hadrosaurus foulkii, 29
Half-life, 73, 83, 99, 190
Hansen, Thor, 122, 134–135
Hawaii, 104
Hawkins, Benjamin Waterhouse, 29–30
Health issues, of dinosaurs, 31
Heezen, Bruce, 53
Helium, 97
Helium-3, 175
Hell Creek Formation, 22–23, 177, 178
Hess, Harry, 40, 53–54, 58
Hess, Victor, 76
Hiatus, 82, 190
Hildebrand, Alan, 139, 142
Holmes, Arthur, 74
Holocene Epoch, 74
Hull, Pincelli, 165–168, 171–173, 184
Hunterian Museum, 21
Hutton, James, 67–70
Hyde, Earl, 100
Hydrogen atoms, 97
Hydrogen bubble chamber, 78–79
Hylaeosaurus, 21
Hypothesis, 2, 27, 190

Ibn al-Haytham, 27
Ichthyosaur, 19
Igneous rock, 70, 190
Iguanodon, 21, 29
Immunizations, 63
Imperial College (London), 175
Imperial measurements, 2
India, 127–128, 169, 173
Indonesia, 104
Impact crater, 190
Impact hypothesis, 103–109, 126, 129, 146–147, 171–173
Impact sites, 90, 92, 131–137, 139–147, 173–176. *See also* Craters
Impact winter, 183
Impact-melt glass, 145–147
Investigating, as scientific practice, 2, 27
Ion, 190
Ionizing radiation, 76, 190
Ionosphere, 98
Iridium
 from around the world, 107, 108
 asteroid impact and, 144, 156, 175
 calculating time of K-T event from, 106
 debate about, 108–109, 126–127
 description of, 85
 method of measuring, 85–86
 in Scaglia rossa limestone, 87
 in Stevns Klint samples, 97
Iron
 measuring, 141
 in rocks, 60–61

Isotope, 73, 190
Italy, 59, 60–65, 81

Jefferson, Thomas, 22–23
Jin Dynasty, 20
JOIDES *Resolution*, 165–167
Journals, 63, 192
Junium, Christopher, 183
Jupiter, 101
Jurassic Period, 35, 72, 74

Keller, Greta, 127–128, 169–171
Krakatoa volcanic eruption (1883), 104, 105
K-T boundary/K-Pg boundary/K-T extinction
  in Apennine Mountains, 60–65, 81
  article on, 107
  debates about, 108–109, 125–128, 168–171
  defined, 190
  discovery of, 64–65
  drilling impact site of, 173–176
  events in, 149–163
  forams in, 61–65, 143
  formation of, 84
  fossil site of, 177–182
  ideas about, 101–102
  impact hypothesis on, 103–109, 126, 129, 146–147, 171–173
  iridium in (*See* Iridium)
  radiometric dating of, 83
  search for impact site, 131–137, 139–147
  season of impact, 182–183
  spherules in, 129–130, 143–147
  in Stevns Klint, 95–97
  supernova theory on, 97–100
  volcanism theory on, 108, 127–128, 169–173

Lamont-Doherty Earth Observatory, 60, 83, 136, 139, 172
Large Body Impacts and Terrestrial Evolution (Conference), 126
Law of Superposition, 35, 82, 190
Lawrence, Ernest, 46, 80
Lawrence Berkeley Laboratory, 99–100
L/B *Myrtle* (boat), 173
Lead, 73
Leprosy, 44
Lewis, Merriwether, 22–23
Limestone. *See* Scaglia rossa limestone
Linear accelerator, 49
Lithosphere, 56, 191
Lituya Bay (Alaska), 111–122
Logic, 27
Longoria, Jose, 142
Los Alamos, New Mexico, 50, 51
The Lost World (Doyle), 30
Lowery, Christopher, 173–175, 184
Lowrie, Bill, 60–65
Lunar and Planetary Institute (LPI), 126
Lunar dust, 89
Lyell, Charles, 25, 70

Magma, 53–54, 56, 57, 191
Magnesium, 129
Magnetic polarity, 54, 56, 191
Magnetic reversals, 54, 191
Magnetometer, 141, 191
Mammals, 161
Manhattan Project, 51
Manson, Iowa, 133–134
Mantle, 55, 56, 191
Marchi, Simone, 176
Mariner (Martian probe), 94

Marl, 143, 191
Mars, 90, 152
Marsh, Othniel, 28–29
Mass extinctions, 98–99, 126, 167–170, 183
Matthews, Drummond, 54
Maurrasse, Florentin, 136
Mayo Clinic, 45, 46
McCay, Winsor, 30
McKee, Chris, 101
Measurements, 2
Megalosaurus, 18, 21
Meghalayan Age, 74
Mesoproterozoic Period, 74
Mesozoic Era, 74
"The Mesozoic Terminated in Boreal Spring" (During), 183
Messinger, Gary, 43, 47–49
Metamorphic rock, 70, 191
Meteor(s), 102, 191
Meteorites, 102, 191
Meteoroids, 102. *See also* Asteroid(s)
Meteorologist, 36, 191
Metric system, 2
Mexican Petroleum Institute, 146
Mexico, 140–142
Michel, Helen V., 99, 105, 107
Microfossils. *See* Foraminifera (forams)
Microtektites, 178, 180
Mid-ocean ridge, 53, 56, 191
Miller, Don, 118
Millner, Mildred "Milly," 58–60, 81, 83–84, 99–100, 142–144
Miocene Epoch, 74
"Missing link," 30
Mississippian Period, 74
Modeling, as scientific practice, 1
Montanari, Sandro, 132, 142–144
Moon landing, 89
Moon rocks, 89
Morgan, Joanna, 175
Mudstone, 135, 136
Muller, Richard, 83, 138
Murchison, Roderick, 72
Museum displays, 29–30
Museum of Natural History (Paris), 23

Nagasaki, Japan, 51
National Academy of Sciences, 126
Natural selection, evolution through, 26, 189
Natural theology, 24, 191
Nature (journal), 183
Nebular clouds, 101
Neoproterozoic Period, 74
Nesvorny, David, 176
Neutron activation, 85–87
Neutrons, 73
New Mexico, 108–109, 127
New Zealand, 107, 108
Nobel Prize, 77–80
North America
  expeditions in, 22–23
  fossils found in, 28–30
  Western Interior Seaway in, 124, 181
North Dakota, 177
Nuclear physics, 46–51
Nuclear test site, 91, 92
Nucleus, 79

Observations, making, 1
Ocean acidification, 167
Ocean Drilling Project, 139–140
Ocean floor, 41–42
Ocean pH, 167

Officer, Charles, 127
Oligocene Epoch, 74
On the Origin of Species (Darwin), 26
"One Hundred Million Years of Geomagnetic Polarity History" (Lowrie and Alvarez), 62
Ordovician Period, 72, 74
Owen, Richard, 18, 21–22, 24–25, 29
Oxygen, 101, 167, 171
Ozone layer, 98

Pacific Ocean floor, 41
PaleoBOND, 177
Paleobotanists, 33
Paleocene Epoch, 74
Paleogene Period. *See* Tertiary Period
Paleontology
  Bone Wars, 28–30
  classification in, 18, 21–22
  defined, 192
  early discoveries, 16–18, 21
  gender discrimination in, 16, 19
  K-Pg fossil site, 177–182
  process of finding fossils, 177
Paleoproterozoic Period, 74
Paleotsunamis, 121, 184
Paleozoic Era, 74
Paleozoologists, 33
Palm forests, 160
Pangaea, 39
Particle(s)
  in cosmic dust, 85–87
  Stokes's Law on, 105–106
  subatomic, 49, 73, 75–77, 78–79, 193
Particle accelerator, 48–49
Parts per billion (ppb), 85
Peer-review process, 63, 192
Peirce, Charles S., 27
PEMEX, 140–142, 146
Penfield, Glen, 141, 142, 146
Pennant, Thomas, 22
Pennsylvanian Period, 74
Penrose Medal, 94
Periods, 74
Permian Period, 72, 74
Petrified wood, 130
pH
  of oceans, 167
  of sediments, 173
Phanerozoic Eon, 74
PhD, 46, 192
Philippines, 104
Photons, 86
Physics, 46, 80, 192
Physiology, 33, 192
Pinney, Anna, 19
Plafker, George, 118
Plankton, 156
Plate tectonics, 53–57, 192
Plesiosaurus, 17, 19
Pliocene Epoch, 74
Plot, Robert, 21
Plutonium, 51
Plutonium-244, 99–100
Polarity, magnetic, 54, 56, 191
Potassium, 129, 130
ppb. *See* Parts per billion (ppb)
Precambrian Era, 74
Princeton University, 30, 40, 53, 58, 127
Principles of Geology (Lyell), 25, 70
Problem-solving, scientific, 1–2, 27, 192
Proceedings of the National Academy of Sciences (journal), 182
Proterozoic Eon, 74
Protons, 73
Pyramids, Egyptian, 81

Qualitative information, 1
Quantitative information, 1
Quartz
  elements in, 129
  shocked, 92, 131-132, 193
Quaternary Period, 72, 74
Questions, asking, 1, 27

Radiation
  dangers of, 138
  defined, 73, 192
  ionizing, 76, 190
Radioactive, 192
Radioactive decay, 73
Radioactive elements, 73
Radiometric dating, 73, 83, 146, 175
Raton Basin (Colorado), 108, 109
Red sandstone, 72
Richards, Mark, 180
Rift, 53, 192
Ring of Fire, 41
Robin, Eric, 145
Rocchia, Robert, 145
Rocks
  in drill cores (*See* Drill cores)
  igneous, 70, 190
  layers of, 35, 69, 71
  measuring age of, 73
  metamorphic, 70, 191
  moon, 89
  sedimentary, 35, 70, 82, 192
Rocky Mountains, 127
Roots of scientific words, 32
Russell, Dale, 97

Sandstone, 135, 136
Sanidine, 130
Scaglia rossa limestone, 60-65, 81, 87
Science (magazine), 107
Scientific arguments, 80
Scientific conferences, 126, 137, 162, 169
Scientific journals, 63, 192
Scientific practices/scientific method
  defined, 192
  development of, 27
  peer-review process and, 63, 192
  tools of, 1-2
Scientific words, roots of, 32
Sculptures, 29-30
Seafloor spreading, 54, 56, 192
Season, of impact, 182-183
Sediment
  in drill cores (*See* Drill cores)
  pH of, 173
  from tsunamis, 120-122, 135-137, 174
Sedimentary rock, 35, 70, 82, 192
Seed-bearing plants, 161
Seiche wave, 155, 180, 192
Seismic waves, 180-181
"A Seismically Induced Onshore
  Surge Deposit at the KPg boundary,
  North Dakota" (DePalma, Smit, and
  Alvarez), 182
Shang dynasty, 20
Shocked quartz, 92, 131-132, 193
Shoemaker, Carolyn, 91
Shoemaker, Gene, 91, 92, 132, 133
Shooting stars, 102
Silica, 130
Silicon, 129
Silurian Period, 72, 74
Silva, Isabella Premoli, 62
Smit, Jan
  articles written by, 182, 184
  at Brazos River site, 135
  iridium found by, 107
  photos of, 145, 177

samples collected by, in Mexico,
  142-144
at Snowbird conferences, 126, 169
spherules discovered by, 129-130
at Tanis site, 179-180
Smithsonian Museum (Washington,
  D.C.), 30
Smithwick, Geraldine, 46, 50, 51-52, 75
Smoke
  from asteroid strike, 158
  from volcanic eruptions, 104-105
Smoking, 63
Smyth, Harriet, 44
Snowbird, Utah, 126, 169
Sodium, 129
Solar flare, 101
Sonar, 40, 193
Spain, 107, 129
Spherules, 129-130, 143-147, 193
Star(s), 97
Star dust. *See* Cosmic dust
Stegosaurus, 28
Steno, Nicolaus, 35
Stevns Klint (Denmark), 95-97
Stokes, Sir George, 105, 106
Stokes's Law, 105-106, 193
Strata, 35, 71, 82, 193
Stratigraphic, 193
Stratigraphic map, 71
Stratigraphic record, unconformity
  in, 82, 193
Subatomic particles, 49, 73, 75-77,
  78-79, 193
Subduction zones, 55, 57, 193
Sulfur, 152, 169, 183
Sulfuric acid, 156
Super solar flare, 101
Supernova, 97-100, 193
Superposition, Law of, 35, 82, 190
Swinburne, Nicola, 142-144
Synopsis of Quadrupeds (Pennant), 22
Syracuse University, 183

Tanis fossil site, 177-182
Tectonic plates, 53-57, 192
Tektites, 89, 144, 146, 153, 193
Tertiary Period, 63-64, 72, 74. *See also*
  K-T boundary
Testing, as scientific practice, 2, 27
Tharp, Marie, 53
Theology, natural, 24, 191
Theory of the Earth (Hutton), 69-70
Tidal waves, 101
Tobacco use, 63
Transition zone, 174
Traps (volcanic flows), 127-128, 169, 173
T-Rex and the Crater of Doom (Alvarez),
  184-186
Triassic Period, 35, 72
Tsunami
  asteroid strike causing, 153-155,
    179, 180
  earthquake causing, 111-117, 119
  paleotsunamis, 121, 144
  as possible cause of K-T extinction,
    101
  sediment from, 120-122, 135-137, 174
  volcanic activity and, 119
  and waves, 119
Tucker, Wallace, 97
Tunguska impact (1908), 103
Tunisia, 169

Ulrich, Howard, 111-117
Ulrich, Sonny, 111-117
Unconformity, 82, 193
Uniformitarianism

vs. catastrophism, 25, 26, 33, 94, 107
  defined, 193
  development of concept of, 25
  evolution in, 26
  plate tectonics in, 53
Units of measure, 2
University of Arizona, 139
University of California, Berkeley, 43,
  44, 46-49, 50-52, 75-77, 84, 108
University of Chicago, 46, 51, 76, 92
University of Manchester, 176-177
University of Southern California, 172
University of Texas, 172, 173, 175
University of Washington, 123, 135, 180
Uplift, 57, 193
Uppsala University, 182
Uranium, 73, 91
US Geological Survey (USGS), 28, 91, 92,
  118, 131, 146

Vine, Fred, 54
Volcanic activity
  Krakatoa volcanic eruption (1883),
    104, 105
  map of, 41
  and shocked quartz, 131-132
  traps, 127-128, 169, 173
  and tsunamis, 119
  and the year without summer (1816),
    104-105
Volcanism theory, 32, 108, 127-128,
  169-173

Waves
  seiche, 155, 180, 192
  seismic, 180-181
  tidal, 101
  tsunami and, 119
  wind and, 119
Wegener, Alfred, 36-39, 53
Wegener, Kurt, 36
Western Ghats (India), 128
Western Interior Seaway, 124, 181
Wiberg, Patricia, 137
Wind wave, 119
Winter, impact, 183
Women. *See* Gender discrimination
Wood, petrified, 130

Yale Peabody Museum, 165
Yale University, 28, 73, 165
Year without summer (1816), 104-105
Yucatán Peninsula, 140-141, 142, 144, 151
Yucca Flats, 91, 92

Zacatecas, Mexico, 142-144
Zanoguera, Antonio Camargo, 142
Zerkle, Aubrey, 183

PHOTO CREDITS

Courtesy of shipseducation.net: 15; courtesy of Wikimedia Commons: 16, 17 (Thomas Webster, top), 17 (bottom), 18, 19, 20, 21 (Biodiversity Heritage Library), 24 (both), 25 (Wellcome Collection), 26, 27 (both), 29 (bottom), 30 (Winsor McCay), 44 (right), 46, 48 (Science Museum, London), 49 (Kebuk awan), 51, 52 (Gku), 55, 59 (F3rn4ndo), 60 (physicalmap.org), 61 (Khruner), 62 (Biodiversity Heritage Library, left), 62 (Hannes Grobe/AWI, right), 76 (top), 78, 79 (U.S. Department of Energy, top), 82 (Patrick Mackie), 85 (Amara), 97 (Judy Schmidt), 99, 102 (bottom)/109/178 (James St. John), 103 (Denys/historicair), 104, 126 (Apollomelos), 128 (CamArchGrad, top), 128 (Nicholas/Nichalp, bottom), 131 (Glen A. Izett, left), 133 (USGS/D. Roddy), 137, 141 (NASA, both), 165 (Bahnfrend), 166 (UCL Mathematical & Physical Sciences), 170 (Lyndsey R. Fox, Stephen Stukins, Tom Hill, and Haydon W. Bailey), 181 (Ron Blakely, Colorado Plateau Geosystems, top), 185 (U.S. Department of Energy); courtesy of the Library of Congress: 23, 29 (top); Flickr, accessed January 19, 2024: 44 (top portrait); Steensma David P, Luis Walter Alvarez: Another "Mayo-Trained Nobel Laureate," Mayo Clinic Proc 2006; 81(2):241-244. Used with permission of Mayo Foundation for Medical Education and Research, all rights reserved 44 (second portrait); courtesy of Hanna Holborn Gray Special Collections Research Center, University of Chicago Library: 44 (third portrait); courtesy of Jan Smit: 44 (bottom portrait), 64, 130 (both), 131 (right), 145, 174 (bottom), 177, 178, 179, 181 (bottom); *Incurable Physician*, personal photo: 45; Lamont-Doherty Earth Observatory and the estate of Marie Tharp: 53 (both); courtesy of Andy Buffington: 75, 77; courtesy of Lawrence Berkeley National Laboratory, © The Regents of the University of California, Lawrence Berkeley National Laboratory: 76 (middle), 86, 107; courtesy of the Division of Medicine and Science, National Museum of American History, Smithsonian Institution: 79 (bottom); courtesy of Cindy Jenson-Elliott: 96, 140; courtesy of New England Historical Society: 105; courtesy of Jody Bourgeois, PhD: 123; courtesy of United States Geological Survey: 124; courtesy of Florentin Maurrasse, PhD: 136; courtesy of Glen Penfield: 142; courtesy of Lowery@ECORD_IODP: 174 (top)